THE WORLD'S GREATEST FIX

THE WORLD'S GREATEST FIX

A History of Nitrogen and Agriculture

G. J. LEIGH

2004

Oxford New York
Auckland Bangkok Buenos Aires Cape Town Chennai
Dar es Salaam Delhi Hong Kong Istanbul Karachi Kolkata
Kuala Lumpur Madrid Melbourne Mexico City Mumbai Nairobi
São Paulo Shanghai Taipei Tokyo Toronto

Published by Oxford University Press, Inc.
198 Madison Avenue, New York, New York 10016

www.oup.com

Library of Congress Cataloging-in-Publication Data
Leigh, G. J.
The world's greatest fix : a history of nitrogen and agriculture /
G. J. Leigh.
p. cm.
Includes bibliographical references
ISBN 0-19-516582-9
1. Nitrogen fertilizers—History. I. Title.
S651 .L55 2004
631.8'4—dc22 2003021366

9 8 7 6 5 4 3 2 1

Printed in the United States of America
on acid-free paper

To Amelia for all her patience and support,
and to Daniel and Hannah, who should never cease to ask why

PREFACE

Most of my professional life, for about thirty years, was concerned with the chemistry of nitrogen fixation. My involvement began in early 1965 when I received a letter from Joseph Chatt asking me whether I would care to apply for a position in the new Unit of Nitrogen Fixation, which by then had just settled at the first of what was then the latest generation of British universities, at Brighton in Sussex. This Unit was funded by the then Agricultural Research Council (ARC), and its remit was to discover the mechanism of biological nitrogen fixation. This had been perplexing scientists for about 100 years.

The intention was to explain how it was possible for bacteria living in temperate or tropical environments to mobilise atmospheric nitrogen when industrially this seemed to require catalysts, temperatures of the order of 400 degrees Centigrade, and pressures of some hundreds of atmospheres. The Unit assembled a multi-disciplinary group of scientists, ranging from inorganic chemists to microbiologists. The Unit developed rapidly, both in size and in the scope of its activities. In time it became a unique establishment where scientists of many different kinds were able to talk to each other, and even to understand one another. Visiting scientists came from all over the world in order to experience the unique atmosphere of the laboratory, the foremost in the world. For 30 years a stream of innovative, fundamental discoveries poured from its researches.

It was only towards the end of this period that the funding authorities began to question the value of this research. In the initial stages, the attitude of the ARC was very relaxed: here they had in the Unit some good chaps doing good work, many acknowledged as world authorities, producing a wide range of new knowledge, from inorganic chemistry to biochemistry, molecular biology, and bacterial physiology. During the 1980s and 1990s this *laissez-faire* attitude changed. The successors to the ARC asked questions about the economic benefits and agricultural applications of our work. It has to be admitted that these were very few, not only from the work in Brighton but also from nitrogen fixation work anywhere else in the world. The Sussex laboratory was wound down and disappeared during the decade of the 1990s.

It goes without saying that I personally regretted this change, and I believe that my opinion is due only in part to my advancing age. In the current world, knowledge is rarely valued for itself, and much more often for its commercial potential. Nevertheless, for nearly 30 years my colleagues and I had the immense privilege of studying a challenging problem with a minimum of bureaucratic interference. During this time, I became aware that we were all members of a long line of investigators that stretched back for thousands of years. Each of us saw the problem of soil fertility, expressed for us as the conundrum of biological nitrogen fixation, in a different way, and each of us added a small brick to the imposing edifice that is modern agricultural science. What I have attempted to show in this book is how human beings have solved the problems relating to soil fertility using imagination, ingenuity, and an understanding of how the world works. I also discovered that the story of this work is not just a dry retailing of scientific facts and discoveries. It also shows those involved as real human beings with all the faults, sometimes considerable, of human beings, including jealousy, greed, and even war. This book is dedicated to all of them: maybe their names are beyond recall, but their work remains, with wise use to the benefit of all humankind.

We do not yet know how biological nitrogen fixation works at the atomic level. I used to think that with the knowledge of nitrogen chemistry accumulated by inorganic chemists, then, once we knew the structure of typical nitrogenase enzymes, we could simply extrapolate from the chemistry to the structure and say: that's how it works. We now have all this information, and it is still not obvious how nitrogenases work. Nature is more complex than we imagined. Nevertheless, I hope that this book will show the general reader how challenging, rewarding, and beautiful is science, and will convince them of the vital importance that even lay people should have informed opinions about the impact of science on society. I also hope that the book will stimulate a new generation of researchers to take up the challenge of understanding the beautiful world of biological and chemical science. The benefits to the individuals involved and to human knowledge will be immense. The economic benefits may not yet be evident or measurable, but they are bound, in the course of time, to be considerable.

ACKNOWLEDGMENTS

I would like to acknowledge the unstinting help of the individuals and institutions listed below. Without their assistance, this book would not have been possible. Any errors in the text are my responsibility alone. I have tried correctly to acknowledge and respect all my sources and copyright obligations, both in the text and in the list below, and I apologise if I have inadvertently ignored or misrepresented anything or anyone.

American Cancer Society
American Geographical Society
Dr. S. Becker, BASF Unternehmensarchiv, Ludwigshafen, Germany
Professor R. A. Berner, Yale University, New Haven CT, USA
Professor O. Bøckmann, Norsk Hydro A/S, Porsgrunn, Norway
Bodleian Library, Oxford, UK
Brighton and Hove Local Studies Library, Brighton, UK
British Association for the Advancement of Science
British Museum, London, UK
Clendening History of Medicine Library, University of Kansas Medical Center, USA
Professor T. P. Culbert, University of Arizona, USA
Professor P. J. Dart, University of Queensland, Rockhampton, QLD, Australia
The Dorset County Museum, Dorchester, UK
Professor R. R. Eady, John Innes Centre, Norwich, UK
Environment Agency, UK, particularly the Peterborough regional office
Prof. W. P. Fehlhammer, Deutsches Museum, Munich, Germany
Dr. G. Fisher, Kemira Growhow (UK) Ltd., Cheshire, UK

Dr. S. Flatebø, Norsk Hydro A/S, Porsgrunn, Norway
Dr. Urs L. Gantenbein, Medizinhistorisches Institut, University of Zurich, Switzerland
Germanisches National Museum, Nuremberg, Germany
Dr. Liwy Grazioso Sierra, UNAM, DF Mexico
Professor B. Herold, Instituto Tecnico Superior, Lisbon, Portugal
Professor A. E., Johnston, Rothamstead Experimental Station, Rothamstead, UK
Dr. P. D. Jones, University of East Anglia, Norwich, UK
Dr. B. Leigh, The British Library, London, UK
Dr. W. Martindale, Askham Bryan College, York, UK
National Galleries of Scotland, Edinburgh, UK
National Portrait Gallery, London, UK
The National Trust for England
Dr. S. E. Nielsen, Haldor Topsøe A/S, Denmark
The Nobel Foundation
Regents of the University of Wisconsin, Madison, USA
The Royal Library, Copenhagen, Denmark
Victoria and Albert Museum, London, UK
Sächsische Landesbibliothek- Staats- und Universitätsbibliothek, Dresden, Germany
Lord Townshend, Marquess of Raynham, Norfolk, UK

CONTENTS

THE WORLD'S GREATEST FIX

CHAPTER 1

Nitrogen Fixation, Agriculture, and the Environment

This book tells the story of how humans have used their ingenuity throughout history to maintain soil fertility and to avoid famine through productive agriculture. The struggle to provide sufficient food has been a preoccupation of humanity since the earliest times. As circumstances have changed and as lifestyles have changed, the way in which the food supply has been ensured has also changed. The story of how different peoples have developed solutions to what is essentially the same problem tells us much about human beings of all kinds and in all ages. It shows us how humans have optimised the opportunities available to them by using the resources, both physical and intellectual, that have been available to them. It shows us the similarity amongst human beings of every era. It also demonstrates how one generation builds upon the knowledge of its predecessors to provide a solution that is appropriate to the new conditions, and it also illustrates the way in which science is gradually and painfully built by generations of researchers in a cooperative undertaking that slowly refines the models of reality used to analyse nature.

Traditionally, agriculturalists have tended to be conservative, and this is very understandable. It is stupid to experiment with questionable new methods if you know that the old techniques work and that not using them will risk a year of famine. The Egyptian and the Britons depicted ploughing with very similar implements in figure 1.1 would probably have shared many ideas on how best to raise crops. A survey of how some ancient civilisations attempted to solve the problems of maintaining soil fertility is given in chapter 2. Many of their techniques are still applied somewhere in the world to this day.

The main focus of this book will be on the story of the essential nutrient nitrogen because nitrogen is often the element whose supply limits the

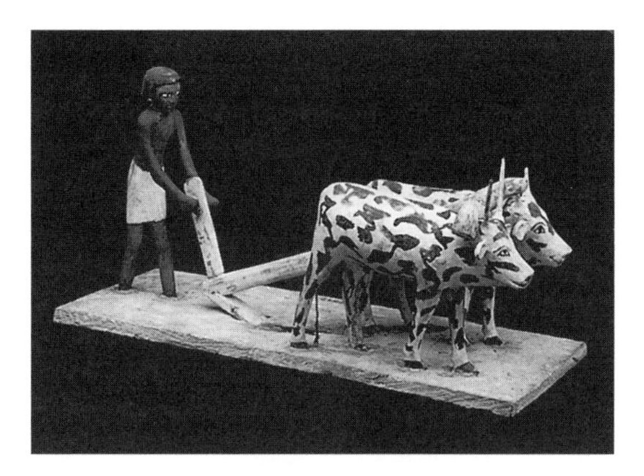

Figure 1.1. (a) An Egyptian ploughman of the Middle Kingdom, 2040–1750 B.C., (© The British Museum, London). (b) A bronze statuette of a Romano-British ploughman from early in the modern era, now in the British Museum, London (© The British Museum, London). (c) A ploughman using a team of oxen in Sussex, in the south of England, in about 1905 (courtesy of Brighton and Hove Libraries, Brighton, Sussex, England).

agricultural productivity of many food systems. Nitrogen is an element that many people know a little about. Nitrogen gas comprises about 80% of Earth's atmosphere, though this was not known 250 years ago, nor would such a statement have made much sense then. Nevertheless, gardeners today are well aware that nitrogen is necessary for good plant growth. Gardening programmes on radio and television often discuss ways of ensuring a good nitrogen supply through proper use of fertilisers. Ammonium nitrate is a common source of nitrogen that is readily available as a fertiliser in many a garden centre or shop (and some people have used it as a cheap explosive for less peaceful purposes!). It is a product of the chemical industry and not, to use the popular term, "organic." Legumes, such as clover, can supply "organic" nitrogen, and the use of fallow periods to allow soils to restore their fertility is dependent upon such plants. This kind of knowledge is useful but hardly complete. How can a plant that must extract nitrogen from the soil as it grows actually enrich the soil with nitrogen, as legumes seem to do? If matters such as these are common knowledge, what can a book about nitrogen tell us? What is nitrogen anyhow? Why should anyone be interested in its history? And what on Earth is the "greatest fix" referred to in the book title?

The answers to some of these questions are very easily found, but to understand the significance of nitrogen for the development of mankind and to understand why the quaint word "fixation" really hides centuries of unfinished human endeavour is to understand a natural process, and this requires much more effort. Nitrogen is an essential requirement for all life as we know it. The need to provide sufficient nitrogen to grow food and support agriculture has governed the rise and fall of civilisations. Some peoples developed sophisticated methods of providing adequate amounts of fertiliser to support agriculture. This went some way in ensuring a reliable and safe food supply, even though, at a time when the only recognised elements were earth, air, fire, and water, as defined by Greek philosophers about 500 B.C. (and similarly by Hindu thinkers), they could not have known the real reason why they did certain things and what all the consequences might be. Nitrogen caused (and still causes) an intense scientific controversy that has lasted two centuries. Nitrogen has been the focus of wars between nations. The discovery of how to convert atmospheric nitrogen to ammonia industrially and economically arguably also made World War I as long and awful as it was. Nevertheless, this discovery also enabled mankind to feed a world population of six thousand million (to avoid confusion between British and U.S. usage of the words billion and trillion, these large numbers have been presented as multiples of a million) and provided the basis of much of the modern chemical industry. It has also led to pollution and sometimes to deaths. Fixed nitrogen can be used as a necessary fertiliser and also as an explosive. It can be used to advantage or it can be misapplied, like much of scientific knowledge. It has made modern industrial civilisation possible. The way in which all this happened is recounted in this book.

Nitrogen, Chemistry, and Fixation

Alchemy and the Origins of Chemical Science

The word "fix," as used in the title of this book, is alchemical in origin. The *Oxford English Dictionary* defines fixation in this sense as "The process of reducing a volatile spirit or essence to a permanent bodily form." It was first recorded in English in 1393: "Do that there be fixation With tempred hetes of the fire." Fixation means the conversion of something volatile and mobile to a solid form. Modern chemists no longer use the word. During their researches of perhaps 1000 years, alchemists invented much of the classical chemical manipulations and even some of the equipment, and they were familiar with distillation, even though they had rather unsatisfactory ways of classifying materials. Thus, they could distil liquids, which passed from the liquid state to a gaseous (vapour) state and then back again when the vapours condensed.

Alchemists have been traditionally represented by outsiders as obsessed and stupid men, though females among the founders of the subject include Mary the Jewess (also known as Miriam the Prophetess), supposedly the sister of Moses and the inventor of the bain-marie, and an Egyptian priestess called Cleopatra (not the queen of that name). These two may well have been real persons, and alchemists were certainly people trying, as best they could, to discover how the world functioned, but their conceptual backgrounds were completely unsuited for their goal. The alchemists were certainly not stupid, though some of their experiments were ill-founded. Though they were often looking to prepare the elixir of life or to transmute base metals to gold, their lasting achievements were more mundane and certainly more generally useful. These included the identification of substances such as alcohol, to which they gave delightful Latin names, including *aqua vitae,* or the water of life. They discovered many common chemicals that are still in use today, such as oil of vitriol (sulfuric acid) and *aqua fortis* (nitric acid). Their ideas about chemical constitution were rooted in the concepts of the four elements: earth, air, fire, and water. Today, we have recognised more than 110 elements, none of them earth, air, fire, or water. Alchemists thought that mercury and sulfur were of profound significance for their quest, though alchemical sulfur and mercury were almost mystical, ideal substances rather than mundane chemicals. We shall meet alchemical nitrogen later in the story. For the alchemists, fixation was simply the process of making a volatile material solid, though this process was often given a deep religious and philosophical significance. The concept of a gas as a state of matter and a solid as another state of matter was rather late in being developed. Apparently, the alchemists did not even recognise the existence of gases.

The alchemical profession was rather insecure, and not only because alchemists produced materials that they did not understand and that are today regarded as positively dangerous. Failure, which was inevitable, could be, on occasion, lethal, often for political reasons. They believed that metals grew in the earth, and exhausted mines are known to have been closed for periods to allow the ore within to multiply. Gold was supposed to be the most highly developed of the metals, the climax of such growth processes. We know today that neither eternal life nor transmutation to gold is achievable, at least not in the ways alchemists attempted, but because their research was expensive, they often sought financial support from people such as rich princes, who clearly hoped to benefit from the discoveries. Alchemists, no less than modern scientists, were keen to find sponsors.[1]

The Emperor Rudolf II of Bohemia (1576–1612) (figure 1.2) was one of many rulers who were interested in alchemy. In fact, he had a broad interest in science and also sponsored the astronomers Johannes Kepler and Tycho Brahé.[2] He employed a team of alchemists with the aim of discovering the elixir of life. One can still see today in Prague the tiny houses in Golden Street where these researchers lived and worked. It is evident either that they were very small people or that they were very cramped in their accommodation, and possibly both. They were familiar with the process that they termed fixation. For example, they knew that liquid mercury (quicksilver) could

Figure 1.2. The Emperor Rudolf II of Bohemia (1552–1612). Rudolf was interested in astrology, astronomy, and alchemy, and ways in which humans interacted with nature. One artist even produced a portrait of Rudolf constructed of an artful assembly of fruit and plants (courtesy of the Victoria and Albert Museum, London).

be rendered solid by heating it in air to about 300 °C (though their work was not quantitative and temperature scales had yet to be defined), at which point it became a red solid that is now recognised to be a mercury oxide. The mercury had reacted with oxygen from the air. At a slightly higher temperature, the oxide decomposes to regenerate the liquid mercury and the gaseous oxygen. Of course, they did not interpret this in modern chemical terms, but they were fascinated by the transition from liquid mercury to the solid and then back again. They described such processes in detail, and they often interpreted them in terms of birth and death, of sexual intercourse, and of death and resurrection. Alchemical writings were meant to be understood only by the initiated, and alchemists used codes and allegories to inform their colleagues of their discoveries.

Rudolf II was seeking gold and long life, but the most significant product of his laboratory is rumoured to have been plum brandy, still popular today in Prague and the Czech Republic, and further afield. Rudolf himself did not suffer failure gladly, and when an alchemist appeared to have run out of convincing ideas for further research he was summarily dispatched. Modern treatment of unsuccessful researchers is also severe, though perhaps not as extreme as Rudolf's. He arranged for his alchemists to be executed on a gilded gallows, which is said to have borne the inscription (in Latin): *Once I knew how to fix mercury and now I am fixed myself.* It is ironic that, 150 years later, Lavoisier used precisely this reversible fixation of mercury to overthrow the phlogiston theory and to usher in the modern era of quantitative chemistry.

The alchemists were not stupid or unintelligent, though they were clearly obsessed. They had not developed the scientific method sufficiently, nor had they an adequate conceptual background to enable them to achieve what they wanted. They could not even frame the correct questions. Eminent scientists such as Sir Isaac Newton and the Hon. Robert Boyle were also interested in alchemy. Since Newton was at one time director of the Royal Mint, it is clear where his interest lay. In fact, Boyle apparently gave what he believed to be the secret of transmutation both to Newton and to the philosopher John Locke. Newton was clearly not the most organised of men, and he later wrote to Locke in 1692 saying that "I feare I have lost ye first and third part out of my pockett." One can only marvel at the gullibility of *The Scyptical Chymist* (the title of Boyle's best-known book) and his colleague (the forgetful professor).[3] To understand the science and technology of a given period, one must also understand how the people of that era thought and acted. Newton and Boyle were not fools, but two of the greatest intellects the world has ever seen. Newton was also intensely religious and studied the Bible in great detail. They were, of course, men of their time.

For example, James IV of Scotland (1473–1513) (figure 1.3) was also interested in alchemy and was something of an experimentalist himself. He once marooned newly born twins with a mute wet-nurse on the Scottish island of Inchkeith in the Firth of Forth in an attempt to discover what was the original language of mankind, which, it was reasoned, the twins would obvi-

Figure 1.3. James IV of Scotland (1473–1513). This painting may be an actual portrait. James improved administrative and judicial procedures in Scotland and encouraged manufacturing and shipbuilding. Reproduced with permission of the National Galleries of Scotland, Edinburgh, Scotland.

ously use between themselves if no other influences were present. Apparently, it turned out to be "goode Hebrew," though the experiment has, as far as I know, not been repeated.[4] This research would be judged crazy today, but at that time it was carried out in a very reasonable fashion to answer what was not then a stupid question. The importance of such researchers was that they were experimentalists, unlike the classical Greeks, who thought and theorised, and thus they laid the foundations of modern experimental science.

There were lots of stories of successful alchemists and transmutations, some of them not adequately rationalised even today. These stories must have arisen from misconceptions or from successfully contrived illusions without any obvious basis in reality. One of the most successful alchemists was Johann Friedrich Böttger,[5] who worked at the court of Augustus the Strong of Saxony (figure 1.4) in Dresden and never produced gold. However, he did discover how to make Chinese-style porcelain, which turned out to be a veritable gold mine for Augustus. Dresden (and Meissen, where Böttger was incarcerated and forced to make porcelain in secret so that no one else would learn the secret of how to do it) was the first European centre of highly prized and expensive porcelain, before that obtainable only from China. The manufacture of porcelain at Meissen became an enormous success. As Shakespeare wrote, "All that glisters is not gold," but, equally, there are things that do not glister that are even more valuable than the yellow metal.

This was the rather unpromising foundation upon which the structure of modern chemistry was erected. The story of how the world gradually came

Figure 1.4. Augustus the Strong, Elector of Saxony (1670–1733). Augustus was a great patron of the arts and sciences and beautified Dresden, making it a centre of European importance; he was also the founder of the Meissen porcelain factory, which used a process discovered by the alchemist Böttger. Augustus was also noted for the large number of his children. Reproduced by kind permission of the Sächsische Landesbibliothek- Staats- und Universitätsbibliothek, Abt. Deutsche Fotothek, Dresden, Germany.

to develop the ideas and the experience of the alchemists and to understand the bases of chemistry, especially the chemistry behind nitrogen fixation, and to develop more secure ways of fertilising the soil, is told in chapters 3 and 4.

All the Nitrogen Chemistry You Need to Know

Chemistry is often regarded as a difficult subject to master, requiring a considerable effort of memory to be a good researcher. However, the basic notation and ideas of chemistry, though at first sight rather obscure, are not difficult to learn. All that is necessary to understand the story told in this book is presented here.[6]

The concept of fixation is easily appreciated in the case of mercury described above, but the problem with nitrogen is a bit more complicated. Nitrogen is element number 7 in the classification of the elements used by chemists, the periodic table. It is designated by the symbol N, but the symbol N standing alone means an isolated nitrogen atom. There are few elements that exist under normal conditions as isolated atoms (the rare gas neon, symbol Ne, is an example), and nitrogen is not one of them. Nitrogen is normally found combined with itself or with other elements, as compounds. The most abundant source of nitrogen on Earth is in the air we breathe. Apart from the oxygen we require simply to function, which constitutes about 20% by volume of normal air, the other major constituent, almost 80% of the volume in air, is nitrogen, but not nitrogen in the form of nitrogen atoms. Both the oxygen and the nitrogen exist in the air as pairs of atoms, designated by chemists as O_2 and N_2, respectively, and there is a particularly strong link, difficult to break, between such pairs of nitrogen atoms. Although conventionally we often refer to these atmospheric constituents simply as oxygen and nitrogen, it is more accurate and informative to refer to them as dioxygen and dinitrogen, and that we shall do for the rest of this book.

Nitrogen (N) is a basic element of all life forms on Earth, along with carbon (symbol C), oxygen (O), sulfur (S), hydrogen (H), and phosphorus (P). This is by no means a complete list, and it is currently believed that about thirty elements are necessary for the proper functioning of human beings and of living things in general. Many of these are trace elements, required only in small amounts, though their absence from the diet can be fatal. However, among the elements required in bulk, those listed above are the lightest (in terms of the weights of individual atoms) and the most important. In living organisms, nitrogen is a principal constituent of amino acids. There are about thirty of these recognised in nature, and humans are incapable of synthesising them in their bodies. They must be obtained ultimately from plant sources. Amino acids are all characterised by having a backbone that is a chain of carbon atoms (sometimes the chain consists only of a single link or C atom) and two groups of atoms attached to that chain, an acid group, represented by -COOH, and an amino group, represented by $-NH_2$ (or H_2N-; the order of presentation is not important, just the atomic ratio). It is not necessary to study chemistry at great length to understand these representations, and they will not appear often in this book, but it is essential to comprehend them in order to grasp the problems we shall discuss. The simplest amino acid is glycine, which we can represent by the formula $(H_2N)CH_2(COOH)$, by which we mean that there is a central carbon atom to which are attached an acid group and an amino group, plus two hydrogen atoms to saturate the bonding requirements of this central carbon. These two groups, the acid group and the amino group, are the basis for the name given to this kind of compound, amino acid. As everybody who eats modern prepared foods and reads the packaging will know, these acids are essential for proper functioning of the human body, though the food manufacturers do not often explain why, and there can be a considerable amount of hyperbole in the promotional literature describing their culinary delights.

The principal function of amino acids in biology is as constituents of proteins. Proteins are formed by what chemists call condensation reactions.

In a typical condensation reaction, an amino -NH_2 group reacts with (or condenses with) an acid -COOH group to eliminate water, H_2O. In this case, the acid provides an O and an H, and the amino group provides a further H. The denuded carbon and nitrogen atoms cannot exist free having lost these atoms, so they saturate themselves again by joining together. This kind of process may be represented by an equation:

$$-NH_2 + -COOH \rightarrow -NH-CO- + H_2O$$

Because each amino acid possesses both an amino group and an acid group, it can take part in two such condensation reactions, one at each end, and the ultimate product is a chain of amino acid residues joined by bonds between carbon and nitrogen such as that shown in the equation above. The general name for such a chain material is a polymer, and there are lots of polymers known, such as polythene and PTFE (Teflon), though these may be of no biological interest whatsoever. Proteins, on the other hand, mediate all kinds of processes within cells, and they are also biological structural building blocks. With the range of available amino acids and the variable chain lengths exhibited by proteins, the variety of possible structures open to proteins is enormous.

That is not the end of the biological requirements for nitrogen. The other major group of biological materials that contains nitrogen is the nucleic acids, of which the most widely known popularly are deoxyribonucleic acid (DNA) and ribonucleic acid (RNA). DNA itself is not a single substance, and it varies considerably in constitution from organism to organism. The nucleic acids ultimately control most aspects of biological function, including the synthesis of the proteins described briefly above and, of course, heredity, through the genes. The nucleic acids are composed of rather complicated chemical compounds called bases, but there are only four employed in DNA and RNA, namely, adenine (designated by the letter A), guanine (G), thymine (T), and cytosine (C). These four bases are arranged in groups along a chain, and the total DNA within a cell constitutes the genome of that cell. One sequence of some thousands of millions of these bases is the recently unraveled human genome, which can be expressed simply as a very long string of various combinations of the letters C, G, A, and T. The genome contains all the information necessary to "construct" a cell and, from it, an organism. The significant thing from our point of view is that all the bases C, G, A, and T contain nitrogen.

Nitrogen is involved at every level of biological function. We breathe an atmosphere that is predominantly dinitrogen, N_2. Dinitrogen is a very unreactive material, and gaseous dinitrogen is therefore generally not available for plants to use in growth. Fixation of nitrogen is the chemical conversion of the gaseous form to a solid or liquid and much more reactive fixed-nitrogen compound in which form plants can now take up the nitrogen.

It is not always possible to ensure that there is enough fixed nitrogen to fulfill the growth requirements of our crops, our animals, and ourselves. Some nitrogen in a fixed form is available in the environment. It may be present as nitrate, represented by chemists as a molecule carrying a negative charge, NO_3^-, as nitrite, also a molecule carrying a negative charge (NO_2^-), or even as ammonia, NH_3. If a plant or an animal has sufficient fixed nitrogen provided to it, then there should be no problem, but a nitrogen deficiency can ultimately lead to death. Crops that have restricted growth due to a lack of nitrogen are termed "nitrogen-limited." Sometimes, other elements are limiting. Gardeners are familiar with phosphorus deficiencies, and copper and potash (essentially a source of the element potassium) are used liberally in fertilisers. However, the major limitation to plant growth in modern intensive farming is often nitrogen availability.

Nitrogen is also the basis of many chemicals of commerce, some of which are produced on an enormous scale. Dinitrogen itself can react with dioxygen in the air under the influence of lightning to produce oxides of nitrogen, often denoted by the formula NO_x. The x in this case means that this is really a mixture of compounds, including N_2O (nitrous oxide), NO (nitric oxide), and NO_2 (nitrogen dioxide). In conversational speech this mixture is often referred to as "NOX," though it is written in various ways. Nitrous oxide is a "greenhouse gas" that is produced naturally by bacterial action. The NO_x dissolves in moisture in the atmosphere to produce a mixture of HNO_2 (nitrous acid) and HNO_3 (nitric acid). These are components of acid rain and are also produced naturally as part of the natural nitrogen cycle (figure 1.5). However, nitric acid is also required in industry as a reagent, and it is manufactured on an enormous scale, nearly 7 million tonnes of 100% acid in the United States alone in 2002. Note that both units, ton and tonne, are in general use throughout the world. One ton (2000 pounds) = 0.907 tonnes (1000 kilogrammes). They are similar but not the same. Nitric acid can be (and was) produced by an industrial process directly parallel to the natural process just described, from dinitrogen and dioxygen. Burning ammonia in dioxygen in the presence of a catalyst now generates the necessary oxides of nitrogen. Often the nitric acid and further ammonia are combined subsequently to form ammonium nitrate. These topics will be discussed further in chapter 5.

Nitrogen Fixation and the Chemical Industry

The Fixation Problem

The problem of providing enough fixed nitrogen if sufficient amounts are not naturally available is simply this: the dinitrogen molecule has between its two nitrogen atoms one of the strongest bonds known to chemistry, and this makes it very unreactive and stable. It is chemically no trivial matter to activate or fix dinitrogen and convert it to a form that is usable by plants and animals. It was not until about 1900 that chemists were able to develop at least three processes to do this on an industrial scale. The most important, and the

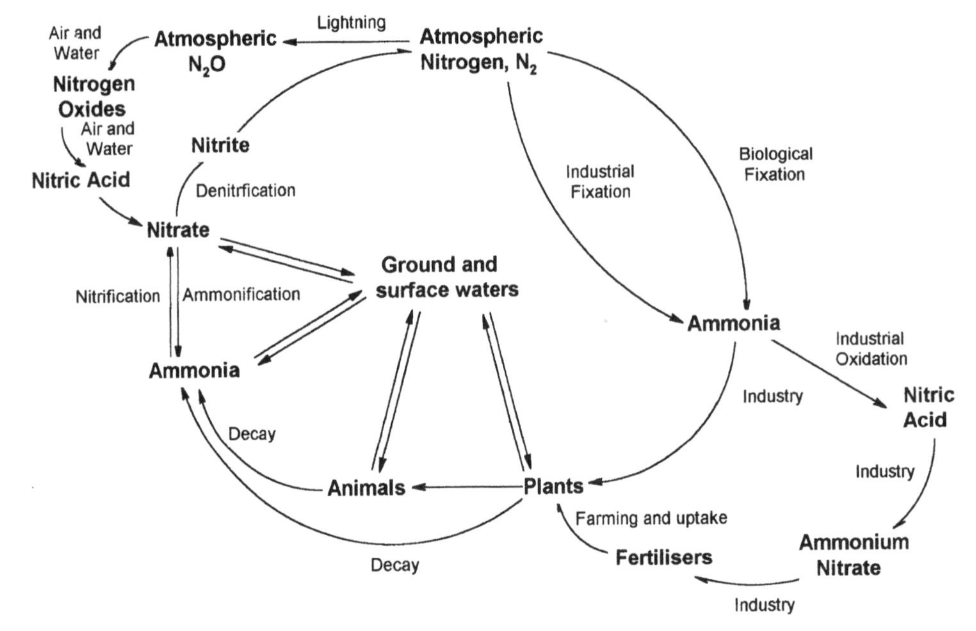

Figure 1.5. An example of a nitrogen cycle. Such cycles may in fact be much more complex than a simple cyclic connection of chemical species. This cycle attempts to show the principal pathways and chemical species that concern nitrogen in the environment. Note that no account is taken of fluxes (that is, the rates of the various transformations and the quantities of substances involved). Such diagrams can be designed to illustrate a number of points—for example, uptake of nitrogen by plants, washout of nitrogen from the soil, and effects of fertiliser application—depending on the intention of the designer.

only one still operative, is generally called the Haber process (or sometimes, and more accurately, the Haber–Bosch process), after one of its principal discoverers, Fritz Haber. No plants or animals can fix nitrogen by themselves unaided. So if this source of atmospheric nitrogen were otherwise unexploitable, Nature would appear to have presented us with an insurmountable barrier to the increase of biological productivity. In fact, Nature has had millions of years to solve this problem, and has done so in ways we have yet fully to understand.

Sir William Crookes' Clarion Call to Chemists

The address of Sir William Crookes (figure 1.6) to the British Association for the Advancement of Science in 1898 is popularly supposed to have brought the nitrogen deficit (the insufficiency of a fixed-nitrogen supply to ensure an adequate level of food production) to the attention of the world scientific community, though it is doubtful whether he was the first to do so. Nevertheless, the end of the nineteenth century certainly was a turning point in the history of intensive agriculture and of industrial chemistry.

Figure 1.6. Sir William Crookes, President of the British Association for the Advancement of Science, 1898. This painting was executed in 1885 by Albert Ludovici and is reproduced by courtesy of The National Portrait Gallery, London. Later pictures of Crookes, made when he was perhaps more in the public eye, portray a much smarter figure, with neatly parted hair and stiffened moustachios pointing up rather than drooping, as in this painting.

Crookes was a man of his time, and some of his ideas may seem unpalatable to many people one hundred years after he expressed them. His address[7] was divided into two parts, one concerning the nitrogen problem, which he analysed in terms of limitations on world wheat supply, and the other dealing with the mysterious psychic world that was at that time linked with such things as X rays and other emanations that were being discovered but not yet understood. Indeed, after his address, he wrote to a friend, a female spiritualist: "The Wheat (curtailed) did not bore, and the psychic part did not shock the audience. The latter part, indeed, got most of the applause. It was all very gratifying."[8] This was not such a quixotic reaction as it may today appear. The argument between evolutionists and believers, materialists and theists, over evolution and creation had been very intense, and John Tyndall, President of the British Association in 1874 and a sceptic, had even proposed that a ward in a hospital be set aside for a scientific study of the efficacy of prayer as a cure for illnesses.[9] I do not know whether the proposal was ever adopted, but the topic is not dead, even today.[10]

Actually, Crookes had a long history of practical involvement with fertilisers, an important constituent of which is, of course, nitrogen.[8] He set up a company to market a process for turning animal waste, chiefly fish, into nitrogenous matter suitable as fertiliser, and for some years, beginning in 1871, he was director of the Native Guano Company at a salary of £200 per year. This company used a process to convert London sewage into saleable manure. Crookes wrote in the *Quarterly Journal of Science*, of which he was a cofounder, that the waste of manurial wealth was as if three million quartern loaves (quartern loaves seem to have weighed four pounds) daily were flushed down the Thames to the sea. In 1892, he demonstrated, though this was not for the first time, that oxygen and nitrogen of the air could combine to form nitrogen oxides if the temperature were high enough.

However, Crookes seems to have come to the conclusion that such efforts were not, in the end, able to provide enough nitrogen-containing fertiliser to ensure a reliable food supply. In his address to the British Association in 1898,[7] he made the point quite graphically: "Are we to go hungry and to know the trial of scarcity? That is the poignant question . . . If bread fails us . . . what are we to do? We are born wheat eaters. Other races, vastly superior to us in numbers, but differing widely in material and intellectual progress, are eaters of Indian corn, rice, millet, and other grains; . . . and it is on this account that the accumulated experience of civilised mankind has set wheat apart as the fit and proper food . . . It is said that when other wheat-exporting countries realise that the [United] States can no longer keep pace with demand, these other countries will extend their areas of cultivation to keep up the supply *pari passu* . . . But will this comfortable and cherished doctrine bear the test of examination?"

Crookes pointed out that a given quantity of wheat that cost 100 shillings in the United Kingdom (£5 in modern currency) cost only 67 shillings in the United States and 54 shillings in Russia. However, prices were bound to rise worldwide as the demand and the price of labour increased, still a familiar litany one hears today, one hundred years later. Crookes forecast that by 1931 wheat lands would no longer be able to supply the bread eaters of the world. One solution was to find a cheap and reliable supply of fertiliser nitrogen to boost agricultural productivity. He issued a clarion call to chemists[7]: "For years past attempts have been made to effect the fixation of atmospheric nitrogen, and some of the processes have met with sufficient partial success to warrant experimentalists in pushing their trials still further; but I think I am right in saying that no process has yet been brought to the notice of the scientific or commercial men which can be considered successful either as regards cost or yield of product. It is possible, by several methods, to fix a certain amount of atmospheric nitrogen; but to the best of my knowledge no process has hitherto converted more than a small amount and this at a cost largely in excess of the present market value of fixed nitrogen."

This plea was shortly to be answered, and in a manner that would change agriculture irreversibly and push back the limits of sustainability very much further (for details, see chapter 5). Modern industry has made it possible to feed populations vastly greater than that of the world of one hundred years ago, and though there are indeed environmental and social problems associated with modern intensive agriculture, the problem of feeding the world population is at least as much political and economic as scientific.

Nitrogen Fixation and Biology

The industrial fixation of nitrogen occurs today on an enormous scale. The basic reaction employed is that of dihydrogen (H_2) and dinitrogen (N_2) to form ammonia (NH_3):

$$N_2 + 3H_2 \rightarrow 2NH_3$$

As stated above, this reaction is known as the Haber, or Haber–Bosch, process. As we shall discuss later, a modern industrial plant can produce about 2000 or more tonnes of ammonia per day, perhaps approaching a million tonnes per year. Such a plant may be a linear assembly of units perhaps 800 metres long. The total annual world industrial production of ammonia is of the order of 100 million tonnes, and it is widely used as a fertiliser, in explosives, and in the manufacture of nitric acid. In nature, a certain amount of nitrogen is fixed as oxides of nitrogen by lightning, but this is hardly a viable, reliable source of sufficient fixed nitrogen. If these were the only natural ways of fixing nitrogen, the world would be very different indeed because there would have been a strict nitrogen limitation on the amount of plant growth for most of Earth's existence.

However, there is a natural biological fixation of nitrogen that can be performed by a series of microorganisms, principally bacteria and some actinomycetes, and these have probably been fixing nitrogen for much of the time that life has existed on Earth. Scientists generally accepted the reality of this fixation only after 1886. It is this biological nitrogen fixation that has maintained a supply of nitrogen suitable for use by other plants and animals, and it is this supply that has enabled many civilisations to develop into the complex societies we now recognise. The scale of this process is of the order of the current industrial fixation, 100 million tonnes per year worldwide, and it underlies the story of human development. For all but the last 150 years, its existence was not generally even suspected, though even earlier Humphrey Davy considered the possibility of biological nitrogen fixation.

We shall discuss later how the existence of biological nitrogen fixation was gradually proved. At this stage, it is sufficient to know that it occurs and that animals or plants cannot effect it. It is a function of microorganisms, and it seems to be a widespread phenomenon. There are free-living bacteria that live in the soil, and these can convert dinitrogen to ammonia, and thence into amino acids. They can be aerobic or anaerobic, and some can fix nitrogen only when they need to, whereas others seem obliged to do so and the fixation processes will then probably require a considerable source of energy in order to operate. The molecular apparatus for fixing nitrogen inside the cell is very complex and is also very oxygen-sensitive, so a variety of methods has been developed by the fixing organisms to protect the enzyme that fixes nitrogen from contact with oxygen. For an aerobic bacterium, this implies that there is some kind of mechanism within the cell to separate air into its two major components. Another group of organisms that can fix nitrogen are the cyanobacteria, formerly called blue-green algae. They form long strings of single cells and can give rise to algal blooms under certain conditions. They achieve separation of dioxygen and dinitrogen by carrying out nitrogen fixation, respiration, and photosynthesis in different cells within the cell chain. Other organisms effect the separation by carrying out fixation and photosynthesis at different times of the day.

Agriculturally speaking, the most important family of bacteria that fixes nitrogen is that of the rhizobia. These are organisms that can live and reproduce in the soil, just as do any other free-living bacteria. However, in that state they do not fix nitrogen. They infect the roots of certain plants through the root hairs, and once inside the roots they change their lifestyle completely. The rhizobia are then contained in nodules attached to the roots, and such nodules have been recognised since at least the sixteenth century. Drawings from as early as 1542 show what must be nodules on the roots of bean plants, although the most famous representation, due to Marcello Malpighi and published in London (figure 1.7), dates from 1675 and 1679 (two printings). Their function was then unknown and could not have been understood at that time. You can find such nodules very easily today by carefully extracting a bean

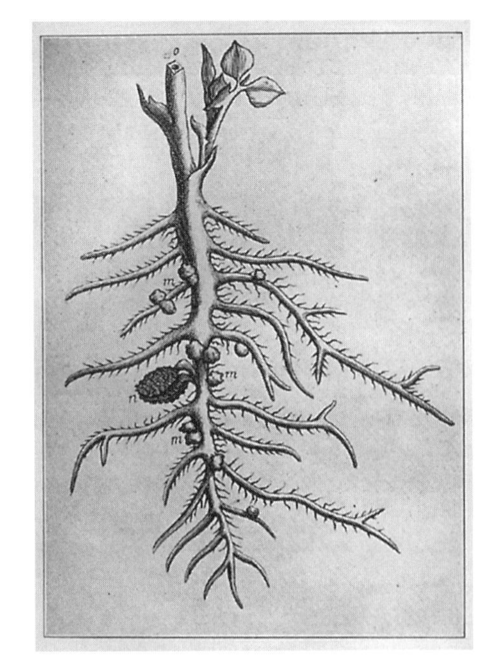

Figure 1.7. Nodules on the root of *Vicia fabia*, from a woodcut by Marcello Malpighi (1628–1694) published in 1679. Of course, Malpighi had no idea of the function of what he was observing. Reproduced from E. B. Fred, I. L. Baldwin, and E. McCoy, *Root Nodule Bacteria and Leguminous Plants*, University of Wisconsin Studies in Science, No. 5, Madison, Wisc., 1932, with permission of the Regents of the University of Wisconsin. A further version of this figure is shown in R. C. Burns and R. W. F. Hardy, *Nitrogen Fixation in Bacteria and Higher Plants*, Molecular Biology, Biochemistry and Biophysics, Volume 21, Springer-Verlag, Berlin, 1975.

plant from the soil or even by inspecting the roots of, say, a clover plant. The clover must be removed from the soil very carefully; otherwise, you risk completely stripping the small nodules and fine root hairs from the roots. The rhizobia can now fix nitrogen within the nodules, and they supply ammonia to the plant in return for the supply of biological energy. This is an example of symbiosis, but the relationship of plant and rhizobium is very specific. The majority of plants that take part in this symbiosis, such as beans and clover, are members of the legume family. Particular strains of legume are infected only by specific strains of rhizobia.

It has long been recognised that legumes (including clovers, peas, and beans) can build up the fertility of soils. Modern agriculturalists know that legumes can supply fixed nitrogen. In less recent times, the knowledge that such plants can enhance soil fertility was widely recognised, though how and why could not be explained or understood. In fact, nitrogen fixers such as the actinomycetes can also fix nitrogen in a symbiotic fashion in conjunction with non-legumes, and other chance associations with rice and cereals are known, though these are not as intimate as the symbiotic relationships. There are also nitrogen-fixing organisms that live on the leaves of some ferns, and these have been exploited as green fertilisers in China. Even some of the recently recognised archaebacteria can fix nitrogen, which implies again that the ability to fix nitrogen has been exploited for much of the time life has existed on Earth. A recent list of recognised nitrogen fixers, including the types mentioned above, can be found in the summary of Eady.[11]

Nitrogen Fixation in the Modern World

Nature knows how to fix nitrogen at very mild temperatures, say between 10 and 60 °C. The usual pressure of dinitrogen of the air is about four-fifths of an atmosphere (the total air pressure, though it varies between narrow limits, is defined under standard conditions to be one atmosphere), corresponding to the 80% dinitrogen content of the air that surrounds us. Nature uses air and water as raw materials, and has had a long time to perfect her methods. Nature's research techniques are evolutionary and not planned. Intelligent humans can also fix nitrogen using the Haber–Bosch process, but to do so we need high pressures (up to 350 atmospheres), high temperatures (up to 400 °C), and a catalyst. Oxygen (more accurately dioxygen) from the air is a poison to the industrial process, which takes place in enormous facilities costing hundreds of millions of dollars (figure 1.8). As raw materials, we prefer dinitrogen separated from the dioxygen of the air by prior treatment, natural gas (principally methane), CH_4, and steam. Nature does all her biological fixation quietly in the soil or on leaf surfaces and at normal temperatures and pressures. What does she know that we still do not know? How far we actually understand what is happening in biological fixation will be discussed in chapter 6.

Despite the considerations above, we have now reached a state where the application of nitrogen fertilisers has become a fundamental part of mod-

Figure 1.8. A modern nitrogen-fixation facility that produces ammonia, essentially from air and natural gas, but must use high temperatures and pressures and also catalysts. This plant is in Ludwigshafen am Rhein, Germany. The size of the plant can be judged from the figure of the man on the bicycle in the foreground. Reproduced by courtesy of BASF AG, Ludwigshafen, Germany.

ern intensive agriculture. The inoculation of soils with rhizobia to enhance biological nitrogen fixation is also an established practice, but of limited application. Farmers in many cultures have long known that unless they treat with manure the soil they exploit for production, productivity rapidly falls. The development of the Haber–Bosch process, which produces abundant supplies of cheap fixed nitrogen—generally marketed as ammonia, nitrate, or urea, or some combination of these—overcomes any limitation imposed by nitrogen availability. For many crops in Western industrialised agriculture, the supply of nitrogen appeared to be limiting, and the more nitrogen you applied to a crop, the greater yield you obtained. Because, not so long ago, farmers were exhorted to produce as much food as possible, no one saw much wrong with this. This was especially true in the United Kingdom after World War II. The yields of, say, wheat seemed to be linearly related to the amount of nitrogen applied. In the middle of the 1990s, perhaps 80 million tonnes of nitrogen fertilisers were being applied worldwide and at rates up to about 400 kilograms (kg) per hectare in some cases. Certainly barley, for example, can absorb up to about 300 kg per hectare, but not all crops can absorb nitrogen so efficiently, and there is also the problem of what happens to the excess of the fertiliser applied. Biological nitrogen fixation can also supply some hundreds of kilograms of fixed nitrogen per hectare per year, but it is known that many nitrogen fixers don't in fact do so if there is enough fixed nitrogen for the organism's purpose already available in the environment. The application of fertiliser nitrogen to the soil is therefore likely to inhibit natural biological fixation.

What happens to the excess of fixed nitrogen applied to the soil but not taken up by the plants is still being determined in detail. The general features are often represented by the nitrogen cycle, illustrated in figure 1.5 above. Presumably, this cycle was once more-or-less closed, with little nitrogen being added or leaving it. However, the widespread use of artificial nitrogen fertilisers has ensured that much more nitrogen is actively moving around the cycle, and also in and out, than there was a century ago.

Generally, nitrogen in the form of ammonia stays adsorbed on the soil particles for periods of years. Ammonia adsorbed on clay is often not directly available to plants, though it can be displaced into solution by positively charged molecules (cations). There are bacteria living in the soil that can gain energy by transforming ammonia, NH_3, to nitrate, NO_3^-. The transformation is called nitrification, and in the soil nitrate behaves rather differently from ammonia. Nitrate is much more soluble in water than ammonia, though this may not be biologically significant, but it is also much more mobile. Some nitrate can be absorbed through the roots of plants and then reduced again to ammonia and used in the synthesis of all the complex biomolecules discussed above. Some can be reduced by other sets of bacteria to give nitrite, NO_2^-, and then in a sequence to nitric oxide (nitrogen(II) oxide), NO, nitrous oxide, N_2O, and sometimes, and finally, back to dinitrogen, N_2. Nitrous oxide is widely recognised as a greenhouse gas, so emission of this gas is of potential harm to the environment.

Nitrate applied directly to the soil as fertiliser can also follow these pathways. However, there is another fate for nitrate that is potentially more worrisome. It can simply wash out of the soil and be carried away in the groundwaters or surface waters. The groundwaters will eventually reach rivers and lakes, though the time it takes for this to happen can be several years. For example, nitrogen applied to soils overlying chalk, as in the south of England, may take twenty years to appear in rivers and streams. This makes it rather difficult to monitor the effects of changing agricultural practices. Once the nitrate is in the surface waters, its influences are various.

Just as plants benefit from the presence of fertiliser nitrogen, so do algae living in lakes, reservoirs, and rivers. Provided the systems are not growth-limited by the absence of other elements, such as phosphorus, the nitrogen can provoke abundant growth of algae, which eventually die. The decaying algae use up the dioxygen dissolved in the water, and this can kill the animal inhabitants of the body of water. The result is an unpleasant and unhealthy decaying mass and an unusable body of water. This process is usually termed eutrophication, though strictly this word means enrichment and applies to the supply of the nitrogen nutrient rather than to the unpleasant results it provokes. However, we shall generally use it in this more widely accepted sense.

The consequences for human and animal health of a grossly enriched nitrate concentration in water supplies are widely feared, but such fears are not always reasonably grounded. In fact, both the World Health Organisation (WHO) and the European Union (EU) lay down upper limits for the amount of nitrate allowed for potable water of 50 milligrams (mg) per litre (L). The principal fear is that large amounts of nitrate can generate stomach cancer and also "blue baby" syndrome. The benefits and dangers of nitrogen in the environment are discussed in more detail in chapter 7.

Nitrogen fixation will be with us, biologically and industrially, for the foreseeable future. Apart from the direct practical problems summarised above, there is still an enormous intellectual challenge facing scientists. This is to discover how it is that Nature can fix nitrogen under such mild conditions, whereas industrially we must employ brute force. There may be much new chemistry that we can learn from the biological process and apply for our own uses. One of the aims of this book is to explain what it is that Nature does, as far as we can understand it. Another is to explain how, through the efforts of generations of dedicated farmers and researchers, we have arrived at our current understanding of biological nitrogen fixation. A third aim is to show how industrial fixation has developed from a laboratory process newly discovered at the beginning of the twentieth century into the impressive and sophisticated procedure we use today. Finally, and perhaps most important, we aim to show how the supply of nitrogen, at first without our understanding and later as a result of scientific insights, has underlain the development of civilisation from the time that humans first became agriculturalists rather than simple hunter-gatherers.

CHAPTER 2

The Development of Agriculture

Maintaining Soil Fertility

The Beginnings of Agriculture

The reason why nitrogen is necessary for all living things was established in chapter 1. We also saw the kind of intellectual and theoretical basis upon which the scientific understanding of that necessity had to be established. We now consider the other side of the nitrogen coin: the kinds of agriculture that were actually practised in various selected parts of the world at different times. It is generally accepted that from the earliest times when human beings first lived in groups, they gathered their food wherever they happened to find it, or else they hunted it. The development of agriculture presumably implied that people became rather less peripatetic because they had to return to their crops, at the least to harvest them. At about the same time, they began to domesticate animals, and this must also have considerably restricted their mobility.[1] Whether they noticed that land became less productive as they repeated cultivation on the same plots year after year can only be surmised. All this occurred about 8000 years ago. There is perhaps one caveat. If people lived in an area where food was plentiful and easy to gather, they would not have needed to wander so far to find their nourishment, but the idea of widespread, static settlements of sedentary hunter-gatherers is not a popular one among the experts.

Once people started to live in cities, there must have been significant changes in agricultural practice, for at least two reasons. First, it is not simple to dismantle a city of permanent buildings every few years and to move it to a more convenient spot for growing crops, so people no longer did that. They tended to stay put. Second, the mere fact that there were people in the cities such as artisans, artists, and priests, who were not productive agriculturally but who needed to be fed, meant that those who continued to raise crops had to produce more food than subsistence farmers who fed only their immediate dependants. The new farmers had to supply food not only to themselves and to their families but also to all the non-agriculturalists living in the cities. They must have encountered problems maintaining soil fertility, and the tactic of moving their crops to new fields every few years could have been used only to a limited extent. The question then arises: How did ancient peoples cope with the problem of decline of soil fertility? A related question that they could not have even considered because of the level of their scientific understanding was: How could they satisfy the requirement for a supply of fixed nitrogen? The selected examples considered thus far may relate only to those few who did solve the problem.

Perhaps satisfactory answers to these questions were never found. In many cases, the size of a population of any kind, animal or plant, is ultimately limited by the food supply. It is quite possible that some populations never solved the fertility problem; they simply did not grow beyond a size that their agricultural methods could support. Famine was a recurring problem in many peoples' "good old days," including the peoples of Western Europe. This is indeed the problem that faces the world's human population today, and, using our current understanding of how plants grow, it is probably not an insoluble one for us. What is evident is that for the total world population to be adequately fed, resorting to "natural" methods will not provide a complete solution. They are not productive enough. In any case, monoculture, the raising of crops of a single plant type in extended fields, is not "natural" whatever the detailed method of cultivation adopted, even if it is considered to be "organic." Nevertheless, it is instructive to enquire how humans have approached the problem of maintaining an adequate food supply in the past.

Agriculture and Past Civilisations

It is often not easy to learn in detail how people carried out their agriculture, what crops they grew, and how they tended them. Nevertheless, some civilisations have left written information. Some models and pictures of farming operations have survived, sometimes for several millennia. Some peoples have even left large statues showing how farmers worked. From such sources, we have been able to gain a reasonable level of understanding of how the Egyptians, the Romans, and, above all, the Chinese worked their fields. However, rarely does this provide all the information we would like. Other civilisations, such as in the Tigris and Euphrates valleys, presumably did what we might do in similar physical environments, so we can make informed guesses about what they did. For yet others, we are still in ignorance. However, it is evident that until a reliable and productive agricultural system was developed, an urban civilisation of any size was impossible. The presence of a reliable and productive agricultural system also implies the existence of an assured supply of plant nutrients. Amongst these nutrients, an important requirement is fixed nitrogen.

Very often, the way we have tried to investigate past civilisations has not been helpful to our understanding of ancient agricultures. Archaeologists like to excavate cities and large monuments. They are attractive targets, relatively easy to identify, and offer governments the possibility of a tourist attraction once they are visible. This kind of research can often attract financial support. In contrast, finding field systems and excavating cultivated plots to determine whether they were flooded or drained, and deducing what crops were grown and when, is much less glamorous. Nevertheless, we are beginning to learn quite a lot about ancient agricultural systems.

Farmers seem to be by nature rather conservative people. Each of the three ploughmen depicted by the models shown in figure 1.1—the farmer on the Sussex Downs in southern England at the turn of the nineteenth century, the Romano-British ploughman (woman?) of the first century A.D., and the Egyptian farmer of 4000 years ago—would have recognised immediately what the others were doing. They were all using similar draught animals, probably oxen. Doubtless, farmers everywhere also realised that soil fertility tended to decrease if a field was used to raise the same crop year after year. Clearly, at some time it must have been noticed that manure from farm animals somehow tended to restore soil fertility. This type of knowledge would have been passed down from generation to generation.

Agriculture before Cities

Even when we can learn what crops were actually grown by farmers in any given area, it is much more difficult to determine what agricultural techniques they employed. Detailed written records are very sparse, and for any civilisation before the Romans they are almost non-existent. It is generally assumed that the development of farming rather than hunting/gathering as a means of providing food necessarily meant that people settled in one place to be near their fields. This is probably true, though matters were never as straightforward as that. There must have been a slow change during which all kinds of mixed lifestyles prevailed. What must also have been controlling factors were the principal food and forage plants that were available and the varieties of domestic animals that were present. The important forage plants seem to have been rice in the East, maize and beans in the Americas, and cereals (wheat and barley) in the Middle East and in Europe.

In addition, a proliferation of cities was not necessary in order to exercise extensive political control. The Mongols built an enormous empire, embracing Central Asia and then south into India, east into China, and west into Europe. It was a larger empire than any constructed until the great European expansions of the eighteenth and nineteenth centuries A.D. The principal means of sustenance of the Mongols seems to have been provided by the horse, which has the happy advantage of being mobile. Though they did build some cities, the settlements of the Mongols were apparently generally rather temporary affairs that were abandoned after some years and moved elsewhere. It is tempting to ascribe this in part to the drop in soil fertility that they must have experienced even if they were aware of the restorative properties of horse dung. Because they used the horses for transport, for fighting, and to some extent for food, they apparently did not keep large herds of horses confined in relatively small areas as one might do with sheep and goats. The beneficial effects of horse manure for soil fertility would not have been easy to recognise or exploit under such conditions. The mobile nature of the Mongols' settlements and their reliance on the horse as a tool of warfare must have made them a peculiarly difficult enemy for the more settled civilisations of the West to tackle. Cities such as Samarkand and Bokhara, which relied upon trade for their eminence rather than on exercising direct political control, were the exception within Central Asia rather than the rule.[2]

Often, civilisations much more recent than the Mongol never quite solved the problem of the conflict between the requirements of a sedentary lifestyle and the need to maintain soil fertility. At the beginning of the twentieth century, the Mohave, Cocopa, and Yuma Indians in the southern United States and Mexico (figure 2.1) apparently still followed their traditional way of living.[3] We can garner a clear idea of how these peoples raised their food from records going back to the era of the Spanish invasion of the Americas and also from the accounts from native American Indians gathered by scientists during the first half of the twentieth century. These peoples grew a wide variety of crops, including maize or corn, tepary and other beans, squash, sunflower, and pigweed (*Chenopodium* sp.), many of which were native to the area. They also still gathered wild fruits and vegetables, which were often abundant, and they hunted game. Their agricultural practices have been described in some detail. They apparently relied generally on the Gila and Colorado rivers overflowing their banks annually to provide them with water, though they also cultivated areas with naturally shallow water tables. People used wide floodplains and riverbanks to cultivate their crops. There is no evidence of widespread canalisation of water or irrigation systems. Although some of the plants that were grown were associated with nitrogen-fixing organisms, there is no evidence that American Indians ever made use of this for fertilising the soil. Their crop rotations were apparently determined by the usual times of germination of a given kind of seed. There was no manuring or artificial fertilising of the soil. The flooding of the river restored fertility, and if, for some reason, the river did not flood, then fertility was not restored. There was no crop rotation in the currently understood sense, and fields were never purposefully left fallow. These American Indians did not use fertilisers of any kind. If fields were not flooded during the wet season (perhaps because of low rainfall), then they might not be planted at all. It was clearly simpler to plant elsewhere where the water supply was adequate.

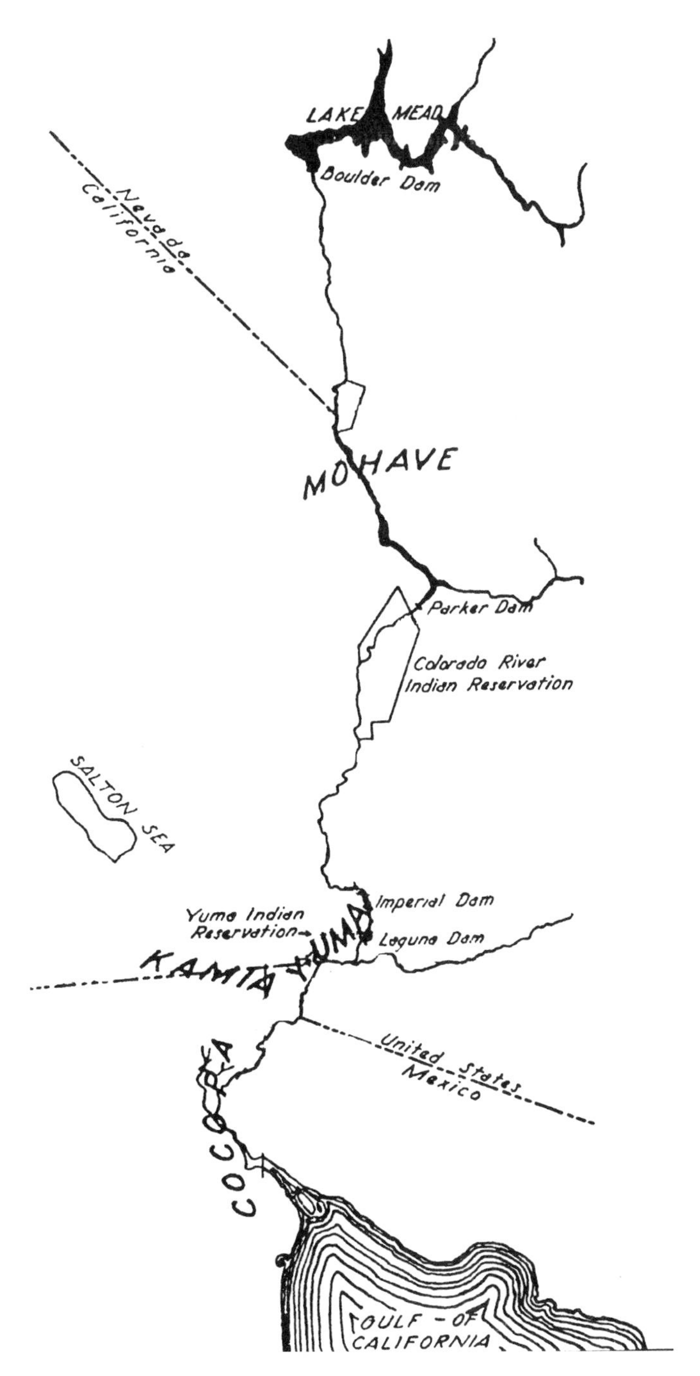

Figure 2.1. The settlement of the Mohave, Cocopa, and Yuma native Americans in the United States and Mexico. This map was adapted from Casteter and Bell, cited in note 3.

In the lower Colorado, this traditional form of agriculture has collapsed only recently. Though similar practices persist elsewhere in Central and South America, the damming of the Colorado effectively prevents any flooding. In fact, so much water is taken from the Colorado, 90% of it for use, especially in agriculture, in the southern United States, that the once mighty river no longer reaches the sea.

The Egyptian Solution: The River Nile as the Foundation of Civilisation

The problem of soil fertility did not often really arise for the ancient Egyptians. The River Nile flooded every year, and when it did so it provided fresh mud and more plant nutrients for the following growing season. The Egyptians were clearly aware of the value of this. They worshipped the Nile and were completely dependent upon it. Of course, it provided the water that all ancient peoples realised was necessary to raise good crops, but what they knew about soil fertility is not obvious. They irrigated their fields using the Nile as a source of water.

The biblical account of Joseph and the sojourn of the Jews in Egypt may well reflect the behaviour of the Nile at that time. The Bible speaks of the seven fat years and the seven lean years that were foretold in Joseph's interpretation of Pharaoh's dream of the fat and lean cattle. The flow of the Nile must have exercised a decisive control on the productivity of the farms in the Egyptian part of the Nile Valley, and this was controlled in turn by the rainfall much farther south, toward the source of the Nile. It would be of interest to look for evidence of prolonged drought in the catchment area of the river at about this time. In any case, the Egyptians (and Joseph) were probably aware that a drought in the south was likely to herald a low flood and a consequent poor harvest in the north. It is now hazarded that the collapse in the years around 1000 B.C. of the Old Kingdom, which flourished in Egypt for about 1000 years, was due to a prolonged drought, maybe associated with a climate change, that drastically reduced the food supply.

Agriculture and the American Civilisations: Pre-Columbian Puzzles

The Americas gave rise to a gamut of advanced civilisations, many of which had already disappeared long before the Europeans began to exploit their colonies in the New World. These people have left an indelible mark on the continent in their monuments and also their writing, much of which has remained almost completely incomprehensible, though the Maya script has recently been deciphered. Now that we can read their inscriptions, we have learnt much about their history, though relatively little about their agriculture.

Those civilisations that survived until the arrival of the Europeans did not find their encounter with the Christian incomers a very happy experience. In their drive to save souls and to accumulate riches, the conquerors regarded all the works of the natives fit only to be damned as works of the Devil. However, there were some more percipient newcomers, generally Spanish monks, who attempted to record details of the civilisations that were being destroyed, so we know a little of the lives of these peoples. The social organisations of the inhabitants of Mesoamerica were very different from those of China or Europe. One might therefore expect their methods of food production also to be different. For example, the Aztecs built a great city, Tenochtitlan (figure 2.2), which seems to have been cleaner, larger,

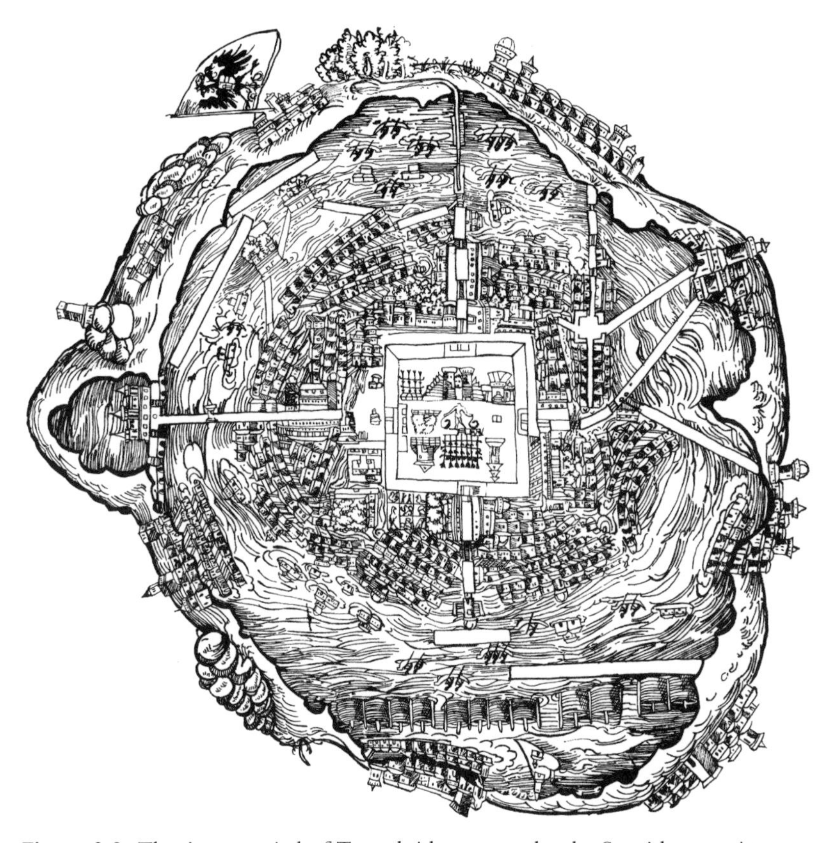

Figure 2.2. The Aztec capital of Tenochtitlan as seen by the Spanish conquistadores. This rather primitive map still clearly indicates the size of the city and the central position of the Templo Mayor, which the Spaniards destroyed and is today being uncovered beside the cathedral in the centre of Mexico City. North is apparently at the top. This map, traditionally attributed to Hernán Cortes, is reproduced, with permission, from Eduardo Matos Moctezuma, *The Great Temple of the Aztecs*, Thames and Hudson, London, 1988, p. 146.

and healthier to live in than the cities then in existence in Europe.[4] Such a city required a large agricultural basis, but we do not know very much about it. One of the most dramatic ways of raising food was to grow it on floating platforms called chinampas and constructed of wood and reeds. Tenochtitlan was situated in the middle of a lake and connected to the mainland by causeways, so this would have enabled the Aztecs to survive even if they were completely blockaded on the lakeshore. Apparently, they raised vegetables on the chinampas, but they, like other similar civilisations, also relied upon maize as a staple, which they used to make tortillas. It is probable that the lake provided their plants with water and minerals, including fixed nitrogen, but we really don't know if or how they ever faced the problem of plant nutrient depletion.

Tenochtitlan was destroyed by the Spaniards after about 1520 and buried under the new colonial Mexico City. It is only now beginning to re-emerge with the construction of the Metro, though some of it, like part of the Great Temple of the Aztecs currently beneath the Metropolitan Cathedral, is unlikely ever to be seen again. The people and their language survived the conquest, as did some of the chinampas in the recreational and market garden area of Mexico City now called Xochimilco, but the agricultural techniques presently employed there are effectively European in origin. However, it is noteworthy that the Chinese also developed a system of floating fields for cultivating areas that might otherwise flood and destroy the growing crops.

Further south in Mexico, and spreading over into Belize, Guatemala, and El Salvador, lived the Mayas (figure 2.3).[5] They have left one of the great enigmas of history. The Mayas, who were not then and are not now a single, uniform people, and their related languages, of which there are more than twenty, survive to this day. The Mayas developed a script that has defied general translation until relatively recently. The Spaniards destroyed all but a few of the manuscripts (codices) they encountered, so most of the written records we have are inscriptions on stone. Details of agricultural practices were not recorded. The Mayas built cities with enormous and mysterious ceremonial centres, though the ordinary people apparently lived in simple huts (palapas) much as many of them still do today. The modern Mayas have been exploited and persecuted in most of the countries in which they exist, and they remain a rather secretive and enclosed people. Their traditions seem to be oral, though some of the ancient religious practices have survived in places, sometimes combined into Christianity.

When the Maya cities were first being generally "discovered" by Europeans, the contrast between their scale and grandeur and the lives of the Mayan peasants caused most Westerners to think that these people could not have been related to those who built the cities. It was a popular idea that the builders might have been the Egyptians, who also constructed great pyramids. As John Lloyd Stephens[6] wrote in his delightful and thought-provoking account of his travels in Yucatan and Central America in about 1840: "America, say historians, was peopled by savages; but savages never reared these structures,

Figure 2.3. The Maya settlement of North and Central America. This map is adapted from one found in R. J. Sharer, *The Ancient Maya*, 5th edition, Stanford University Press, Stanford, Calif., 1994, p. 21.

savages never carved these stones . . . standing as they do in the depths of a tropical forest, silent and solemn, strange in design . . . excellent in sculpture, rich in ornament." Stephens was one of the first explorers to say that the Mayas he encountered on his explorations were indeed the descendants of those who had built the cities.

Of course, the conundrum posed by the Mayas of today and by the silent cities was more than the questions of who built them and why. The Mayas of today still practice what is termed swidden agriculture. This is a method that involves cutting and burning the jungle to provide an area for cultivation. When the productivity of the cultivated area begins to decline after a few years, the ground is abandoned and another plot prepared. The abandoned area, left to itself, gradually reverts to jungle. It takes about twenty years for the exhausted plot to regain its essential nutrients. What seems evident is that if the modern Mayas still use today the methods to support themselves that were used when the cities were alive and populated, then there must have been too few people to support the urban population, let alone to construct the cities in the first place (figure 2.4).

When the Spaniards arrived in Yucatan after 1511, most of the cities, with the exception of places such as Tulum on the Caribbean coast, had long been deserted. The most important Maya period came to an end in the ninth

Figure 2.4. Inca peasants working the land in the sixteenth century. This picture was taken from Felipe Poma Guaman de Ayala, *El Primer Nueva Corónica y Buen Gobierno*, 1613–1615, reproduced with permission of the Royal Danish Library, Copenhagen. This way of working was widespread throughout Mesoamerica. A photograph of Maya peasants working a field in Guatemala can be found in *Arqueol. Mex.*, 5(25), 57 (1977) and is very similar to the earlier picture shown here. In the absence of draught animals and tractors, the land is worked much as it was in classical times, often involving planting seeds individually in holes made with sticks.

century A.D., after perhaps 1000 years of a productive and highly sophisticated existence. The Spaniards subjugated the Mayas, though resistance continued in parts of what is today Guatemala until 1696. The Mayas were exploited most brutally for at least 200 years. No one really could explain why these cities had been built and then, so mysteriously, had been abandoned. The answer accepted until relatively recently was that swidden agriculture could support only a limited population and that this population had simply grown until the food supply could not keep pace. When this happened, people just starved, and they did so in such large numbers that the old cities became deserted. After all, this is a phenomenon that is well-recognised to occur amongst animals, rather less in control of their environment than humans, and it is sometimes described in terms of the balance of nature. However, in the case with the Mayas, the situation might have been more complex.

The Continuing Mystery of the Mayas

Even today, the true extent of the Maya civilisation has not been defined because new cities are being discovered, sometimes, but not always, buried in the jungle. When Stephens (together with the British artist Frederick Catherwood) made his journeys of discovery, he spent his time in Yucatan and the adjacent areas. He was interested in "piedras viejas" (old stones) that were known to the Mayas of that time, about 1840. He saw how the Mayas lived and how they supported themselves, but the principal interest was always the archaeological remains (figure 2.5). This has generally been the case. Questions concerning social organisation, density of population, and agricultural methods were not asked. Indeed, these were probably questions that could not be answered at the time. Certainly, the extent of Maya culture, covering the tropical jungles of the Peten, the dry scrub jungle of Yucatan, areas including Belize and Honduras, and the Western Highlands of southern Mexico and stretching into Guatemala, was not appreciated. Only now are these matters being considered and debated.

What seems evident is that there was a spectacular collapse of population throughout the Maya area between about A.D. 800 and A.D. 850. This collapse was not evident everywhere and did not happen throughout the Maya region at exactly the same time. However, it is clear that after about A.D. 800 no more stelae were erected. Stelae are ceremonial columns inscribed with glyphs (Mayan script) describing the achievements of the local rulers, and they carry dates in the old Maya style, based upon the calendrical cycles that were common throughout Mesoamerica. Most of the ceremonial sites were abandoned, though some were used to some degree for perhaps another two centuries. The

Figure 2.5. Stephens and Catherwood exploring the ruins of Tulum, reproduced with permission from J. L. Stephens, *Incidents of Travel in Yucatan*, volume II, Harper and Brothers, New York, 1841, reprinted by Dover, New York, 1963, p. 265. These two explorers were instrumental in bringing the ruined and forgotten cities of the Maya to the attention of people in Europe and the United States. Stephens' books are still a delight, and Catherwood's engravings are particularly fine. Stephens was one of the first to insist that these cities were built by the ancestors of the Maya he met on his travels and not produced by exotic incomers such as Egyptian immigrants to Central America. See note 6 for more details.

common people lived in essentially temporary structures that were erected on more substantial platforms, and it is clear that relatively few of these were used after the collapse.

What provoked this collapse? Since the final decipherment of the Maya script, which started in the 1950s, our understanding of Maya society has undergone a profound change. The decipherment itself is a romantic story.[7] For at least 400 years, Westerners have puzzled over the script, and some very imaginative solutions were proposed. The only clear decipherment concerned the dates on the stelae, but what they referred to was unknown. Until relatively recently, it was generally believed that these glyphs were ideograms, so that, like Chinese script, they could be read in any language and used by speakers of any language, just as Japanese people can read and understand Chinese inscriptions. No one took the trouble to interpret the script in terms of the Maya languages. This misconception had old roots. One of the few Spanish incomers not bent entirely on pillage and extirpating the Maya religion (at least after his initial visit) was Fray Diego de Landa, who actually recorded the Maya "alphabet" after consulting local inhabitants.[8] Since there really is no Maya alphabet, his consultation must have been rather difficult. For example, the people who were asked by Landa questions such as how they wrote the letter "L" (pronounced "ellay" in Spanish) apparently responded with a glyph that denoted a word that sounded rather like the Spanish pronunciation of the letter L. This exercise was easy, though completely misleading. What generally arose from researches over the centuries was the myth of the Maya rulers as philosopher kings who looked at the stars, invented the concept of mathematical zero, built observatories and immense temples though they had no draught animals (and incidentally no dung to fertilise their fields) and no metal tools, and who survived on the kind of native swidden agriculture still seen in some Maya areas of Mexico today. These kings wrote in a script supposedly understood only by the enlightened few, though most of their written material and many of their stone inscriptions were destroyed with the coming of Christian "civilisation" from Europe, having been designated as works of the Devil. As Landa records: "We found a great number of books in these letters, but since they contained nothing but superstitions and falsehoods of the devil we burned them all."

A Soviet epigrapher named Yuri V. Knorosov finally found the key to the Maya script. He was isolated from the academic community of Maya scholars in Great Britain and the United States, who had not heard of him. He worked alone in Leningrad, nowadays again Saint Petersburg, and was not at that time allowed to visit the countries where the original inscriptions were to be found. In about 1952, he recognised that the script was essentially syllabic and that the syllables were in a Maya language, which is still current. It took a considerable time for the academic community to accept that this self-trained outsider had succeeded where generations of experts had failed. The consequence of this discovery has been enormous. At a very low level, it was shown that an inscribed pot found in a grave was not labeled with some sophisticated incantation but with the word "chocolate." It could have come originally from someone's kitchen! More important, the Maya states were finally revealed from the inscriptions on the stelae as groups of warlike entities continually fighting, exchanging prisoners and booty, and sacrificing their captives. So much for philosopher kings!

This raises many more possibilities to account for the collapse of the Maya civilisation, such as war, social unrest, disease, and famine. Before we try to discuss these possibilities, we must also consider other questions: What was the social organisation of the Maya state, and who produced the food? What kind of agriculture was exploited before the collapse? What was the population density before and after the collapse? Was the collapse a unique event, or was it merely the culmination of a continuing series of events?

The population densities that were assumed until relatively recently were those inferred from observation of current population densities and distributions. However, perhaps the population before the collapse averaged as much as 200 persons per square kilometre.[9] This estimate was arrived at by surveying given areas to determine the densities of structural remains that probably supported residential buildings and then multiplying these densities by an assigned family size. The population varied considerably from place to place and in the highlands of Guatemala was different from that in the lowlands of Mexico. Maximum densities reached perhaps 800 per square kilometre, and the total population must have numbered several million. After the civilisation collapsed, the density was reduced, perhaps by about 70%, and some areas were completely depopulated. The estimates for the period after the collapse are based upon current observed densities.

Maya society is generally assumed to have been divided into two major classes, the elite and the non-elite. This is obviously not a very informative classification and does not tell us anything about social relationships within and between the two groups. It is evident that some cities subsisted upon commerce, such as the production and sale of salt, and others, by implication, must have been concerned primarily with agriculture. How the two kinds of city might have interacted is not clear. There may have been a large peasantry as in China, and there probably was no large slave population comparable to that of Rome. There might have been a mixture of serfs and freemen as in mediaeval Europe, but we just don't know. Certainly, there are slaves portrayed in Maya mural paintings. Landa's account of Maya life gives the impression that the majority of the population was more akin to serfs than to anything else, and these serfs not only fed the whole population but also built the major ceremonial sites. The extent to which this was a myopic transposition of European practices onto the appearances of an alien culture is not clear.

If only the swidden, slash-and-burn agriculture was used, the patterns of land use show that considerable areas of the country were not exploited for food production, and consequently the local densities must have been much higher than the figures quoted above. Once fertility had been reduced by a period of exploitation, the fields were allowed to lie fallow for some years to regenerate. The food crops raised included beans, chilies, cacao, avocado, and papaya, and root crops such as manioc and sweet potatoes. The beans are, as we shall learn, nitrogen-fixing legumes. Apparently, they also grew these plants in kitchen gardens in an intensive way, fertilising them with domestic wastes. More recent evidence from aerial surveys and satellites points to the existence of canals and raised fields in many areas, the kind of constructions that were simply not considered by earlier theorists. Whether there was widespread exploitation of permanent fields is still a matter for debate.

Some recent work has centred on the "bajos," low-lying areas of seasonal swampland that could be cultivated once the water table had receded in the dry season. Seeds were not broadcast but planted in individual holes made laboriously with a stick. Such practices seem to have been common throughout the Americas. It is a matter of conjecture as to whether the canals were used for draining or for irrigation, and in fact both uses may have been exploited in different areas. Nevertheless, the mere existence of a widespread network of canals confirms the formidable nature of the social system that utilised them. What seems to follow is that if these bajos were indeed used for extensive agriculture, then the annual flooding must have also helped with fertilisation of the soil. It seems unlikely that a civilisation without large flocks of domesticated animals and without machinery of any but the most primitive kind would have developed other techniques of fertilisation.

All this seems to confirm that the Maya civilisation before the final catastrophic collapse was highly organised and probably highly stratified. The simple rationalisation for the collapse based upon the inadequacy of swidden agriculture cannot be the whole story. The evidence really does suggest that a large proportion of the population had simply died, and that swidden agriculture supported the survivors alone. Those who lived in the Maya area after the collapse appear to have been mainly the descendants of those who lived in the older social structures rather than conquerors or incomers.

Apparently, the Maya population grew steadily throughout the so-called Classical period, reaching its apogee around A.D. 800. We cannot know whether this rise was steady or punctuated. We also know that there was constant warfare between city-states throughout this period. Life could not have been serene and settled for any of the people involved. The need to find ever more resources to defend cities and attack neighbours, to build monuments, to produce more food, and to support ever more persons who consumed food but never grew any must have placed considerable stress on the society. Whatever the ultimate final cause of the collapse, what happened seems to be that the society simply became limited by its environmental possibilities. A prolonged period of drought, for which considerable evidence is beginning to accumulate, may have been involved. Perhaps disease was also influential, though there is little evidence to confirm this. However sophisticated their agriculture, it does not seem to have been robust enough to fulfill the requirements of Maya society. Since the collapse and the population decline, the Mayas seem by and large to have forgotten the glories and achievements of their predecessors and have eked out marginal existences throughout the Maya lands. For centuries they have been enslaved, bullied, and persecuted by the European incomers and their descendants. Yet their ancestors built one of the truly great civilisations, comparable to any contemporary society anywhere in the world.

Agriculture and Eastern Civilisations, the Chinese Solution

The Chinese situation has been quite different from anything experienced in the Americas or by the Egyptians. There seems to have been a large, settled peasant population from the earliest times for which there is any reliable evidence. How they learned to support their large population is not apparent. Despite all their skill and application, famine was far from unknown in China, and the rulers of the state have always been aware of the danger of hunger. Indeed, the food supply seems often to have exercised a critical control upon population and the stability of society in many different circumstances.

The sophistication of the farming methods in the Far East before European ideas had any great impact upon practice is well-described in a quaint book that appeared at the beginning of the twentieth century. Written by F. H. King, D.Sc., somewhat misleadingly entitled *Farmers of Forty Centuries or Permanent Agriculture in China, Korea and Japan*, and published by Mrs. F. H. King in Madison, Wisconsin, probably a relative of the author, it is really a description of agriculture in these countries of about 1900, and indeed of several of the preceding centuries.[10] Doctor King was obviously a great admirer of the agriculturalists of an area that many Westerners of that era would have dismissed as primitive. According to Dr. King, the United States at that time could provide more than twenty acres of land to support each member of its population, whereas China, Japan, and Korea had had only two acres per capita for the last 3000 years, and one of those two acres was not cultivable. There must be even less land per capita available now, one hundred years later. Doctor King also remarks that the use of imported mineral fertilisers, as in Western Europe and the eastern United States, had never been possible in these Eastern countries. Nevertheless, 500 million people were then being maintained, chiefly on the production from an area smaller than the improved farmlands of the United States. How did they achieve this miracle of production throughout perhaps forty centuries?

Choosing his words rather carefully, Dr. King says that: "the Mongolian races have held all such [sewage] wastes, both urban and rural, and many others which we ignore, sacred to agriculture, applying them to their fields." They also used transported canal mud to maintain soil fertility. Further, he states that ". . . centuries of practice had taught Far East farmers that [leguminous plants] are essential to enduring fertility . . . Just before, or immediately after the rice crop is harvested, fields are often sowed to clover (*Astragalus sinicus*) which is allowed to grow until near the next [rice] transplanting time." The clover was then ploughed in or harvested, stored, fermented, and finally applied to the fields. In fact, the use of "green fertilisers" in the form of plants such as soya beans and adzuki beans has been widespread in China since Zhou times, perhaps since about 1000 B.C. Doctor King stated that: ". . . these old-world farmers whom we regard as ignorant . . . have long included legumes [such as clover] in their crop rotations, regarding them as indispensable." They also practiced intercropping, whereby as many as three different crops maturing at different times of the year were sown on the same field and in adjacent rows.

Doctor King had a high respect for these agriculturalists. He also remarks presciently that the application of mineral fertilisers imported into Europe and the United States could not continue indefinitely, presumably because the supply was limited, though that he doesn't say. The solution to that particular problem, the industrial fixation of nitrogen, was already under way, though Dr. King could not have been aware of it. It will be discussed later, in chapter 5. Clearly, the "Mongolian" farmers had discovered how to solve their nitrogen-deficiency problem, and with some sophistication. How and when they came upon this solution is not at all obvious, but the fertilisers they used were many and various and had been used for centuries.

What is clear is that China has had a relatively settled and highly organised social and political system for millennia. Continuity may have been aided by the early invention of the ideogram system, which ensured that any learned person could read and understand what was written, regardless of his native language or dialect.[11] The borders of the empire had changed through time, and the emperors came from different races and families as a result of war and internal disputes, so life was often far from peaceful. A summary of dynasties and related information is presented in table 2.1, and the competing factions during one especially disturbed time, the Warring States period, in about 260 B.C., are shown in the map in figure 2.6. However, one of the constants of Chinese history seems to have been the realisation that without assured food supplies, a highly controlled, hierarchical, and urban-dominated society could not be stable. Such a food supply relied upon a productive and satisfied peasantry, for China, unlike Europe, had never developed reliance upon slavery, and also, unlike Europe, the consumption of meat from domesticated animals, perhaps apart from the pig, never formed a major part of the diet. In feudal societies, as in the earlier days of the empire, in which the peasants owed a duty of work to their feudal masters, this could only be achieved by ensuring that the peasants were not exploited too rapaciously. Chinese history seems to have been punctuated by a series of land reforms that became necessary as individual landowners exploited and appropriated their neighbours' lands, often in perfectly legal fashion. The Chinese peasant fed the state and also supplied the manpower for major works such as canals, irrigation systems, and defensive walls.

Table 2.1
A summary of the ages of Chinese history up to about 1650, derived from Y. Cotterell and A. Cotterell, *The Early Civilization of China*, Book Club Associates, London, 1975.

	Period covered	Principal dynasties	Important other phases
Beginning of historical period	1500–256 B.C.	Shang dynasty, 1500–1027 B.C. Zhou dynasty, 1027–256 B.C.	Warring states, 481–221 B.C.
Unification	221 B.C.–A.D. 220	Ch'in dynasty, 221–207 B.C. Early Han dynasty, 202 B.C.–A.D. 9 Later Han dynasty, 25–220	Hsin dynasty (Wang Mang), A.D. 9–3
Disunity	221–589	Three Kingdoms, 221–265 Six dynasties, 265–587 Tartar Partition, 317–589	
Reunification	581–906	Sui dynasty, 581–618 T'ang dynasty, 618–906	
Reunification	907–1279	The Two Sungs, 960–1279	Mongol invasion, Yuan dynasty, 1279–1368
Recovery	1368–1644	Ming dynasty	

Land redistribution was seen as a means of assuring the support of the peasant class for the central government and also controlling the landed gentry. A land distribution was generally followed by the gradual assimilation of the peasants' fields into large estates, and the whole redistribution process would begin again. Nevertheless, the central government was able to use the labour of the peasantry to indulge in enormous irrigation projects that benefitted the whole empire, as well as others more widely appreciated outside China, such as the construction of the Great Wall.

China is an immense country covering many climatic areas. The structure of the land was clearly well-suited to the development of a sustainable and productive agriculture, though "sustainability" seems to be a modern term. It is estimated that the population of the Chinese empire was 100 million even at the beginning of the modern era. The Chinese archaeological record goes back to about 1500 B.C., but Chinese civilisation is at least 5000 years old. The Yangshao culture of around 5000 B.C. was already partly agricultural and apparently raised millet. Our first hard information concerning agricultural practices stems from the Shang dynasty (1500–1027 B.C.). These people were deeply concerned with agriculture, and one of their divinities was called "he who rules the millet." This was in the ancient heartland of China, and figure 2.7 shows the extent and position of the Shang empire during this period.

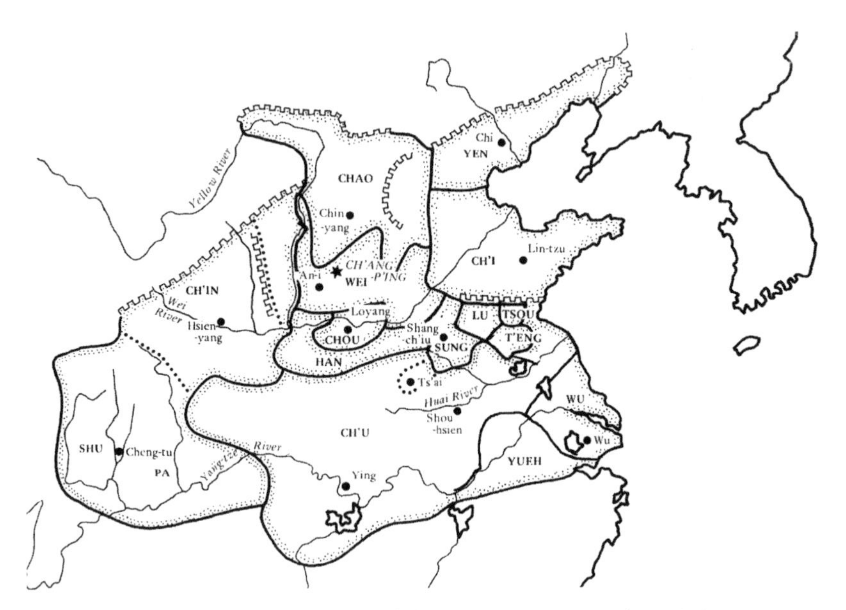

Figure 2.6. China during the period of the Warring States, at about 260 B.C., adapted from Y. Y. Cotterell and A. Cotterell, *The Early Civilization of China*, Book Club Associates, London, 1975, p. 40. The state boundaries are shown by the firm lines, and the castellations denote fortifications. By this time, there were several canals, some of which were hundreds of kilometres in length.

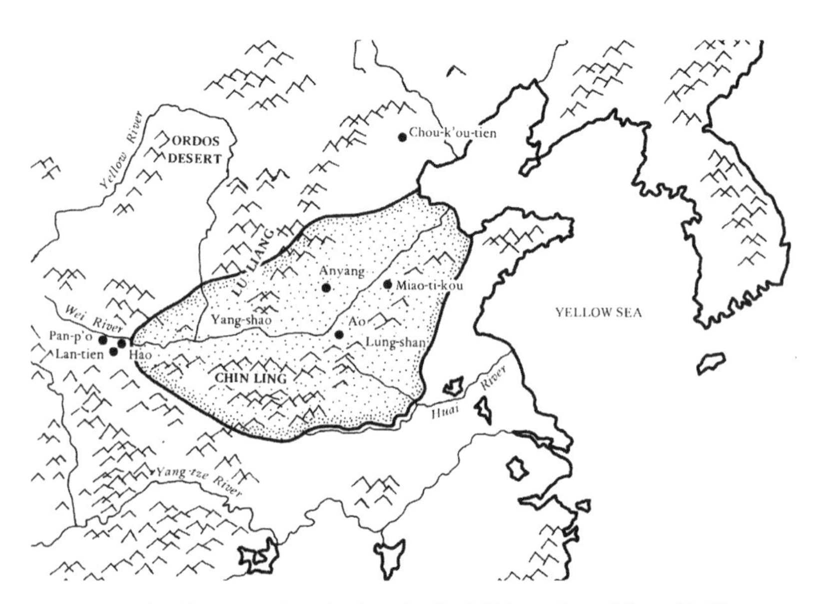

Figure 2.7. The Shang empire, the heartland of China, adapted from Y. Y. Cotterell and A. Cotterell, *The Early Civilization of China*, Book Club Associates, London, 1975, p. 17.

South of the area of the Shang empire, in the valley of the West River, tropical conditions prevail and the growing season is twelve months long. The floodplain of the river is covered by silt hundreds of metres thick. This provides a good source of plant nutrients. Several plantings can reach maturity within a single year. The principal crop was originally rice. In the Yangtse valley, further north, the growing season is nine months, and the plants that can be grown are different. Finally, in the farthest north (the valley of the river Wei), cereals such as wheat and barley are produced. The soil here is composed of loess, wind-blown sand, that can reach thicknesses of many hundreds of metres, and the loess is a repository of minerals and water. A reliable supply of water was early on seen as necessary for continuous, productive agriculture, and the Chinese developed methods for water management, conservation, and irrigation from very early times. They understood enough hydraulics and land management to maintain soil productivity and fertility, and the Chinese peasants were able to pass such knowledge to their descendants. This was at a time when elsewhere there were only faint notions even of the desirability of doing these things.

The written records speak of the equal or equitable field system. This seems to have been based upon a feudal division of land, the origins of which go back at least to the last millennium B.C. Land was apparently first allocated to individual families in the legendary period of the Sage Kings, about 2300–2200 B.C. (see table 2.1 for a list of ruling dynasties). The size of the allocations depended on family needs and the quality of the soil. There are written descriptions of land allotments dating from about 500 B.C. (the Warring States period). The central government of China seems to have claimed the ultimate right to dispose of land. The Zhou dynasty succeeded the Shang period and lasted until 256 B.C. Society was feudal, and the peasantry owed a duty of labour to their lords, but they were still, formally at least, free men and women. The population grew, and rice cultivation spread north. During the Classical Age, 770–221 B.C., the Ch'in reformed the state and also undertook enormous water conservancy works. The emperor Wang Mang carried out an early and important land reform during the first century A.D. Noting that the rich landowners were getting richer and that the poor farmers were losing their holdings, he designated the land of the empire as "the emperor's land," and families with large holdings were ordered to pass their surplus land to those who had none. The peasants were given land to own privately, though because it could be bought and sold it tended gradually to accumulate in fewer and fewer hands. However, the necessity of the support of the peasants for the burgeoning towns and cities was well-recognised. Consequently, during the Han period (A.D. 25–220), the land was nationalised and again redistributed. The emperor Hsiao-wen-ti redistributed the land in the year A.D. 471, apparently assuring the farmers of security of tenure for their lifetimes. By imperial decree, he sent "commissioners to the provinces and commanderies who will, in co-operation with local officials, allot the land of the Empire on a lifetime basis." In A.D. 590, there was another land distribution, by emperor

Sui wen-ti, who decreed that all soldiers should be attached to a prefecture and cultivate the land like peasants. The emperor T'ang T'ai-tsung in about A.D. 630 also enforced the equitable field system. This happened again about a century later. Even later, during the Sung dynasty, there was land reform, and farmers were given state loans against the security of their harvests.

During the later Han period, a characteristic pattern of agriculture emerged. In the north, there were long, narrow fields of wheat and millet on terraces winding around loess hillsides. In the south, there were flooded terraces and paddies. Water conservancy was generally managed on a village basis. Rice cultivation extended to the north of the Yangtse River in the first century A.D. In about 100 B.C., Chao Kuo, an adviser to the emperor Han Wu-ti, devised a system of ploughing that aided planting, irrigation, and soil renewal. He recommended splitting fields into three narrow furrows. This had the advantage of leaving the peaks between the furrows fallow for a season, and during this period the soil regained some fertility. In the following season, the furrows were displaced, concentrating the seed and the water elsewhere and allowing more effective irrigation and weed control. Poor soils were treated with organic wastes, ashes, manures, human wastes, and river silt or canal mud. All these practices were grounded in empirical observation, though scientific analysis on a formal basis was already beginning.

Even then, as today, the cities acted as magnets to attract the overworked peasantry with the promise of an easier life. The population soared, as did agricultural productivity. It is reported that the Empire in A.D. 1124 comprised 20,882,258 households and that the state had a population of more than 100 million persons. In 1270, the population of Hangchow alone was more than one million. Marco Polo recounted in about 1300 how night soil and sewage was collected and then sold to market gardeners. The use of manure was already recognised by Western European farmers at this time, but it does not seem likely that their agriculture, which placed a heavier emphasis on animals and meat than did that of the Chinese, was dependent upon organised refuse collection. Agricultural productivity in Western Europe was much lower than in China.

The Chinese peasants also used complex crop rotations. Simple slash and burn, which implies exploiting soil for a few years until its fertility has declined so much as to render it not worth using and then moving elsewhere, was not widely practiced after about A.D. 0. In fact, fallowing was hardly an option, so great was the demand for food. Crop rotations were well-established by the Han period, and green manures were in widespread use in the Zhou period. Consequently, continuous cropping was possible from early times. For example, to precede millet, the following crops were recommended in order of preference: green gram beans, adzuki beans, cucurbits, cannabis (!) or hemp, sesame, rape, and soybean. The presence of so many legumes in this list is striking. However, there were apparently hundreds of recognised rotations, and it is not easy to summarise them, though most included cereals, legumes, and an oil crop. As in much later European rotations, some ran over several years. Intercropping was also common, using plants such as broad beans.

Clearly, the use of legumes has played a fundamental role in maintaining soil fertility in China for the best part of 3000 years. Apart from the earliest periods, the Chinese farmer was aware of the value of legumes with their (unrecognised) associated nitrogen-fixing systems. Soybeans were cultivated by the seventh century B.C., and from Zhou times. Other beans were also grown, and the green bean was used in rotations as a fertilising agent.

Rice is often seen by Westerners to be the characteristic food of China, and indeed of the whole Far East. Its cultivation is highly productive if the correct resources of water and manual labour are available. It seems probable that rice was domesticated throughout a large swathe of Southeast Asia in very early times, though this is a matter of some controversy. What is apparent in this context is that it was first grown in southern China, where it was also first domesticated. It may have been brought there from India, though authorities differ on this. Although the traditional image of the paddy field is characteristic of the way in which much rice is grown, rice can be grown under wet or dry conditions, and the many different varieties have very different qualities. In particular, some varieties reach maturity within two months of planting, whereas others can take much longer. The advantages of the former are obvious, and appropriate wet-rice cultivation can maintain higher population densities than any other comparable farming system. Probably not much dry rice has been grown in China, and certainly little wet rice north of the Yangtse Valley.

The rice paddies have been manured and carefully tended by hand since early times. Human sewage has always been used as a plant fertiliser, but the Chinese were much more sophisticated than that statement would imply. For example, one authority describes the preparation of the rice seed bed in the following terms: in the autumn, plough it deeply, then cover it with plant waste, which should then be burnt and the ashes ploughed in; in the spring, plough again and spread a manure of hemp water derived from rotted and fermented hemp (night soil is not recommended as it burns the roots of the young seedlings); finally spread chaff and compost on the bed (the best fertiliser for these purposes is a mixture of burned compost, singed pigs' bristles, and rotted coarse bran). After this, the seed may be broadcast.

After this preparation, the rice seedlings were planted out, but the paddies were again intensively prepared and fertilised with river mud, burned compost, hemp, bean cake, and other media, and sometimes green manures were recommended. By the sixteenth century, some authorities were recommending application of potash and oil cake. It was generally recommended to burn the stubble after harvest and to plough it in.

These techniques must have been developed as a consequence of meticulous observation. Their application clearly required painstaking and continuous labour. This may be one reason why China's rulers have generally tried to ensure enough political and social stability to allow the peasantry to work the fields so intensively. However, it is also clear that the agricultural methods employed (though perhaps not necessarily understood at the time in Western scientific terms) can be seen to be very sensible. Much of what the

Chinese did was what a modern Western farmer might do, though in a rather different way. Much of what they did can be interpreted in terms of the need to maintain the supply of fixed nitrogen and minerals.

Agriculture and the Romans

The Romans were great agriculturalists. Like the Chinese, but unlike the Mayas, they left detailed accounts of how to run farms and how to grow crops. Unlike the Chinese, the Romans did not rely on a free peasantry to provide their food. Roman society was a slave society, and the estate owner used slaves or he paid wages to freemen. Also unlike the Chinese, the Roman estate owner, who might well have been a woman rather than a man, was not interested in food production per se. He was interested in profit and pleasure.

There are several accounts of Roman agriculture, many of which offer fascinating insights into the attitudes of the ruling classes of Rome. They are very detailed and describe all that one might wish to know about farming. There are good English translations of the several Roman texts that refer to agriculture. These texts are of considerable interest, as much for their social comment as for their technical information. However, some of the content should not be taken at apparent face value. For example, though we may consider the defining factor of a legume to be something connected with fixing nitrogen, that clearly was not what the Romans had in mind when they used the word. A noted authority, Columella, lists legumes as including the bean, the lentil, the pea, the cow-pea, the chick-pea, hemp, millet, panic grass (otherwise known as Italian millet), sesame, lupin, flax, barley, medic clover (which is probably not a true clover, but related to alfalfa), fenugreek, and vetches of various kinds. Pliny's list is similar, and his descriptions are rather naive: "The bean is the only one of the leguminous plants that has a single stem . . . the lupin also has one but it does not stand up straight." The defining quality of a legume appears to be that it was pulled and not cut when harvested—that the crop was gathered by hand. The *Oxford English Dictionary* ascribes the origin of the English word to the Latin *legere*, meaning to gather. Presumably, legumes were strictly plants that produced a harvest that was gathered by hand and not dug up. It did not include wheat, which was cut, nor apparently figs and olives, though it is not obvious why not.

The Roman Authorities on Agriculture

The Romans were great gatherers of information. Their literary method seems to have been to list as many questions as occurred to them and to read, digest, and regurgitate the writings of earlier authors. To this end, they consulted Greek authors as well as Roman. One of the earliest was Theophrastus, who was born in 370 B.C. and died at about the age of 85.[12] He was a contemporary of Plato and Aristotle and wrote a treatise entitled *Enquiry into Plants*. In the hallowed Aristotelean fashion, this was mainly a classification of all kinds of plants, including their uses, with some discussion of how they are best grown. Theophrastus apparently talks about legumes, though his classification of legumes seems to be that also used by Roman writers. Though not nearly as detailed as his Roman successors, he makes it quite clear that both pot herbs and field crops require manure, and he notes that animal (including human) faeces and urine are of value. The enriching of poor soils by legumes does not seem to have been realised.

The writers most often cited for information on Roman agriculture are listed below. Marcus Portius Cato[13] lived from 234 B.C. until 149 B.C. He was a soldier, politician, and orator and an authority on law and agriculture. Though only a fraction of his writings has been preserved, he is one of the authorities most quoted by later writers. Marcus Terentius Varro[13, 14] (116–27 B.C.) was also a politician, and a devoted student. He was renowned as a *vir Romanorum erudissimus*, a most learned man of the Romans. He is reckoned to have written some 600 books, of which only six survive, three of them on agriculture. He wrote *Res Rusticae* (*On Agriculture*, though the Latin expression seems to be plural), quoted here, when he was nearly eighty and knew that he would shortly die: "My eightieth year admonishes me to gather up my pack before I set forth from life." C. Pliny Secundus, or Pliny the Elder (figure 2.8), was born in A.D. 23 and died in A.D. 79, asphyxiated while observing the great eruption of Vesuvius. His epic, *Naturalis Historia* (*Natural History*), covered every possible aspect of the subject in a work of thirty-eight volumes.[15] Volumes XVI to XIX deal with growing crops of various kinds, though Pliny doesn't go into as much detail as some later writers. Finally, Lucius Junius Moderatus Columella[16] was apparently born in what is now Spain, in Gades (Cadiz), in the first century A.D. He wrote a treatise, *Res Rustica* (*On Agriculture*, but the Latin expression seems to be singular). He based his work on a long list of authorities, including those cited above and also Greek authors, and his is probably the most systematic and comprehensive of all the Roman treatises on agriculture. Pliny also used him, in turn, as an authority.

Pliny's *Natural History* seems to have been an effort to record all human knowledge. It is the earliest encyclopaedia still extant. However, Pliny does not discuss farming in as much detail as, say, Columella. For Pliny, natural history did not include the handling of slaves or the construction of farm buildings, whereas these are treated at length by Columella. Nevertheless, in some places he says more than do some other writers. He is perhaps less critical than others. For example, he discusses the various kinds of soil and the crop plants most appropriate to each. Whether he believed that soil is always necessary is an open question because ". . . some varieties of vine . . . draw nourishment from frosts and clouds." Like the other writers, he also notes things that have an uncanny modern resonance: "In the district of Larisa in Thessaly the emptying of a lake has lowered the temperature of the district, and olives which used to grow there

Figure 2.8. Pliny the Elder, writing in the library. This engraving was published in 1721, and it must be totally a product of the imagination of the artist, possibly one Caspar Gottschling. Pliny died in the first century A.D. during an eruption of Vesuvius. Reproduced by courtesy of the Clendening History of Medicine Library, University of Kansas Medical Center.

before have disappeared, also the vines have begun to be nipped, which did not occur before; while on the other hand the city of Aenos, since the river Maritza was brought near to it, has experienced an increase in warmth." It is possible that drainage works, at which the Romans were particularly skilled, had resulted in local climate changes that were not foreseen.

The How and Why of Roman Agriculture

The order of discussion of topics below is that used by Columella. He addressed his writings to one Publius Silvinus, probably a neighbour, about whom little is known. The work opens with the comment that the leading men of the state were complaining about the deleterious effects of weather and soil exhaustion on agricultural productivity. Columella does not accept that this rather familiar complaint is the fault of anyone but the farmers and the decline of public morals. He decries the life of the city, where ". . . we gaze in astonished admiration at the posturings of effeminate males, because they counterfeit by their womanish motions a sex which nature has denied to men." Later, he says, ". . . what is more to be wondered at, the training schools for the most contemptible of vices—the seasoning of food to promote gluttony and the more extravagant serving of courses, and dressers of the head and hair . . . yet without tillers of the soil it is obvious that mankind can neither subsist nor be fed." Columella was saddened by the change of standards and ideals. He believed that all civilised life depends ultimately on a sound agriculture and food production, and these require hard physical work. He laments: ". . . yesterday's morals and strenuous methods of living are out of tune with our present extravagance and devotion to pleasure." This is a familiar cry, echoing down the centuries!

How to Select Your Estate

A typical owner might try to develop an estate as shown in figure 2.9.[17] The first problem for the aspiring estate owner is to select the right area and the easiest kind of soil where you should purchase or build your farm. This provokes Columella to long discussions about soil type. Pliny covers similar ground, so to speak, and advises checking a soil by taste, though it is not exactly clear what he means by this: "A soil with the taste of perfume will be the best soil." Soils can also be improved, and he ranges through the Roman Empire to find the best methods: "There is another method [to improve the soil] discovered by the provinces of Britain and those of Gaul, the method of feeding the earth by means of itself, and the kind of soil called marl . . . There had previously been two kinds of marl but recently . . . a larger number have begun to be worked . . . Another variety of white marl is chalk used for cleaning silver [!] . . . chiefly used in Britain." Marl also requires fertiliser and dung to assist it in soil improvement. Some farmers also used ashes, often the ashes of burnt stable dung, nowadays recognised as a good source of potassium. Varro also discusses the different soils and the crops appropriate to them, but his discourse is lightened by other discussions, such as the value of cabbage to set you up for a real Roman orgy.

How to Run Your Estate

Columella suggests a very hands-on approach to running an estate. Not for him the life of an absentee landlord. He worries about the licence that slaves and overseers will exercise if they are left too much to their own devices: "But when climate is moderately healthful and the soil moderately good, a man's personal supervision never fails to yield a better return from his

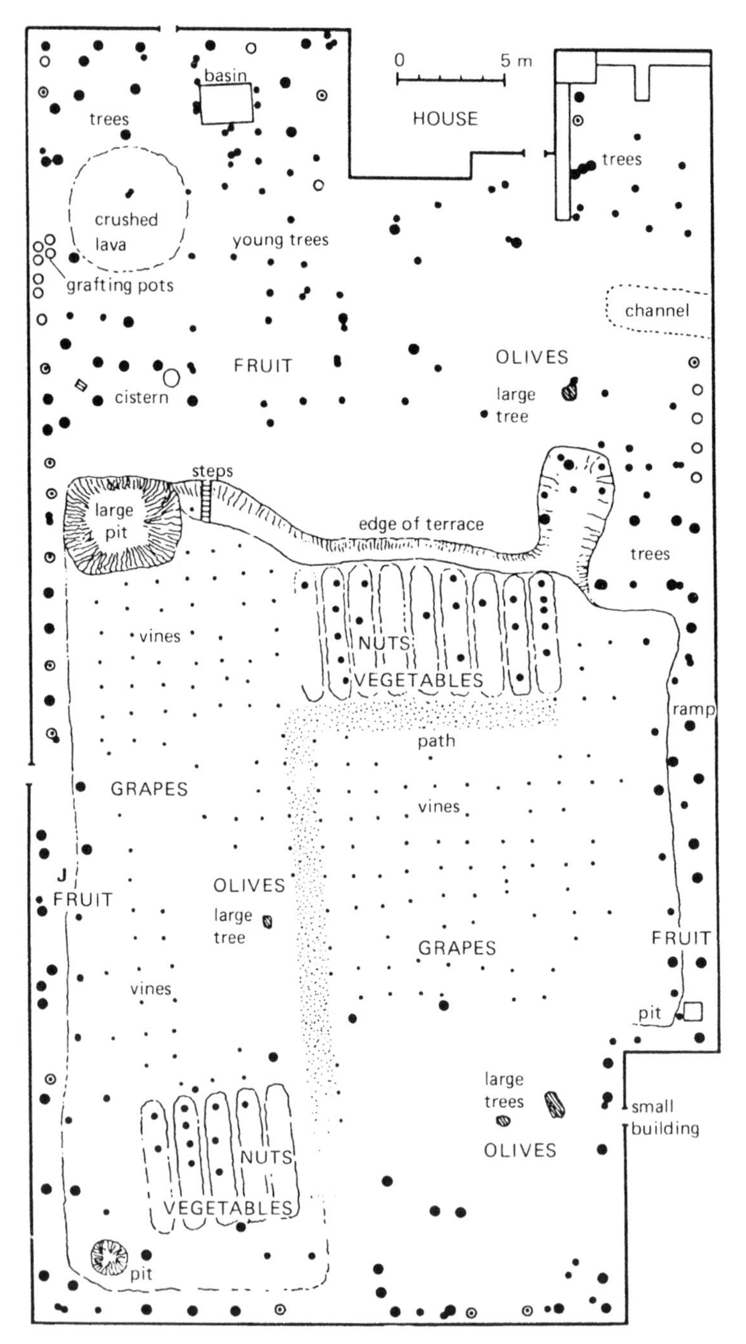

Figure 2.9. Layout of a Roman estate. This sketch represents the garden at Pompeii of the House of the ship *Europa*, so called because of a drawing of a ship scratched on a wall. This diagram is based upon actual excavations and is reproduced with permission from Kevin Greene, *Archaeology of the Roman Economy*, Batsford, London, 1986, p. 97.

land than does that of a tenant . . . or . . . overseer." The overseer should be a slave who is a proven hard worker and not physically attractive because such people are likely to be lazy. Nevertheless, the overseer should be given a woman companion and warned not to become too intimate with other members of the household or with outsiders. Some authorities even describe the amounts of food that are reasonable to give to the farm workers, depending upon the demands of their daily labours. The well-being of the labourers was a prime consideration, even though it was expressed within the standards of the time. Some 100 years earlier, Pliny had remarked: "Nowadays those agricultural tasks are performed by slaves with fettered ankles and by the hands of malefactors with branded faces . . . and we forsooth are surprised that we do not get the same profits from the labour of slave gangs." No details of these farming arrangements are too trivial to consider, lest they affect productivity and profit! Pliny adds: "Good farming is essential, but superlatively good farming spells ruin." This is because it is not economic to take too many pains to produce a crop, and this also has a sad, modern ring. The variety of plants grown in a typical Roman garden is depicted in figure 2.9.

Varro's discussion is rather more foreign to modern ideas: ". . . now I turn to the means by which the land is tilled. Some divide these . . . into three: (the) articulate, the inarticulate, and the mute; the articulate comprising the slaves, the inarticulate comprising the cattle, and the mute comprising the vehicles . . . it is more profitable to work unwholesome land with hired hands than with slaves." Nevertheless, "Slaves should be neither cowed nor high-spirited. They ought to have men over them who know how to read and write and have some little education."

How to Build Your Farm

Columella next describes how the farm buildings should be constructed and what buildings are required. Varro also briefly talks about farm buildings: ". . . their summer dining rooms should face the cool east and their winter dining rooms the west." Amongst Columella's requirements are "at least two ponds, one to serve for geese and cattle, the other in which we may soak lupines, elm-withes and other things . . . There should also be two manure-pits, one to receive fresh and keep it for a year and a second from which the old is hauled." After that, real agricultural techniques are addressed. Columella even tells the intending country dweller to run his (plough) furrows cross-wise to the slope. Although this is modern practice, to stop soil erosion, the reason he adduces is to save the oxen from the repeated toil of ploughing uphill. Informed though the Romans were, it is sometimes unwise to read too much into their recommendations.

Later discussion rejects the idea that soil can be permanently exhausted, that ". . . the earth, the mother of all things, like womankind now worn out with old age, is incapable of bearing offspring . . . But, on the contrary, the soil, whether abandoned deliberately or by chance, is cultivated anew, it repays the farmer with heavy interest for its periods of idleness." This shows clearly that the Roman farmer must have appreciated the value of allowing soil to lie fallow, though whether he or she was keen to forego the income that fallow periods must have necessitated seems rather unlikely.

What to Grow on Your Estate, and Why

The list of plants that the aspiring farmer may grow is then discussed, and this includes the pulses or legumes. Lupins were esteemed first among the legumes: ". . . of all crops that are sown it is most beneficial to the land . . . it flourishes even in exhausted soil . . . it is good fodder for cattle during winter . . . it serves to ward off famine [for humans] if years of crop failure come upon them." As for beans, probably kidney beans, they are best sown on manured soil, but "some people . . . think that in cultivated land this same plant takes the place of manure," which Columella doesn't seem to believe. There must have been a lot of empirical experience to provoke this opinion, but Columella ascribes the beneficial effects of beans to their using less nutrients than other crops. (Varro, on the other hand, recommends ploughing-in young beans as green manure.) No legume is less hurtful to land than the chickpea, whereas hemp (not a legume in modern terms) requires a rich, manured, well-watered soil. Pliny discusses the various uses of legumes for human and animal food in some detail. He also esteemed them highly. Echoing the Roman definition of legume, Pliny quotes Cato to the effect that cereal land is manured by grain, lupin, beans, and vetches, but chickpea scorches cereal land because it is pulled up by the roots: "The most extensively used plant is the lupin, as it is shared by men and hoofed quadrupeds in common." Its seed need not be covered by soil, it flowers and buds three times, it follows the sun (presumably like a sun flower), it provides its own manure, it can be used as an additive to cattle feed for strength and size, and it can be used as a stomach poultice for children. Columella encouraged farmers to use lucerne: "But of those that please us, the medic clover is the choicest because it dungs the land." Pliny also describes the value of lucerne for fodder, though he calls it a grain rather than a legume, as one would today.

In the same context, Varro says that it is not good practice to plant every kind of crop on rich soil. In fact: "It is better to plant in thinner soil those crops that do not need much nutriment, such as clover and the legumes, except the chick pea, which is also a legume, as are those plants which are pulled from the ground and are called legumes from the fact that they are gathered."

Columella quotes more ancient authorities (Saserna) to the effect that a group of legumes (as we would today recognise them) really do fertilise the soil, and then he adds that: "I have no doubt, nor yet as to vetch when it is sown for fodder, provided, however, that (lupin) after being cut green is to be followed immediately by the plough and the ploughshare . . . for this takes the place of manure." Some Roman authorities were clearly aware that other peoples used beans as manure by ploughing them in. Other legumes (Roman style), such as flax and poppies, were said to burn the soil, and then the only remedy is to use manure to recover fertility.

The Value of Manure

Columella's discussion of manure that follows is quite detailed. There are three kinds of manure, produced by birds, by humans, and by cattle, apparently in this order of efficacy. However, not all bird dung is equally good, and it seems that most dung improves with ageing. Pigeon dung is best of all. Human urine is very good for young plants, and old urine is excellent for fruit trees. Varro quotes an older authority (Cassius) to the effect that ". . . the dung of pigeons is the best of these [dungs], because it has the most heat and causes the ground to ferment." Given a choice, he would select the dung of thrushes and blackbirds, but otherwise human excrement, and then various kinds in a series, with horse dung rated lowest. He states that: "The farmer should make a dung-hill . . . They state that if an oak stake is driven into the middle of it no serpent will breed there." Presumably, anyone who had learned otherwise had not survived to report the fact!

Cattle dung is in third place, though because Columella apparently includes asses and sheep under the general heading of cattle it is probable that in Roman times cattle meant more than cows. Indeed, the *Oxford English Dictionary* makes clear that the word "cattle" is related to the words "capital" and "chattel" and in English before about A.D. 1500 it could mean property and animals. Even as livestock, the word seems to have included sheep, camels, pigs, and horses, and only lately has the word been restricted to bovines.

Cato is much briefer in his recommendations. He describes which crops to sow for fodder and which for food. He recommends dividing the dung into two lots and to use half for the forage crops. Place a quarter in trenches around olives. Finally, ". . . save the last fourth for the meadows, and when most needed, as the west wind is blowing, haul it in the dark of the moon." For old vineyards, he recommends sowing clover and applying ". . . around the roots . . . manure, straw, grape dregs, or anything of that sort." Lupins, beans, and vetch fertilise the land, and he advises on the making of compost. As an aside, he is particularly keen on cabbage, which surpasses all other vegetables in medicinal value, and appears to be an efficacious purgative!

Apart from the manures detailed above, Columella recommends the use of oil lees, ashes and cinders, and cut lupin plants, which "provide the strength of the best manure." When even plant and animal manures are not available, then all kinds of vegetable waste can be composted and rotted in pits in the ground. In fact, composting for a year will not reduce the strength of the manure, but it will kill the seeds of weeds that it might contain. All this sounds very impressive, but it is based upon experience rather than scientific understanding. For example, the farmer is instructed to spread the manure in piles of moderate size while the moon is waning and then to broadcast it before sowing. Too much manure is as harmful as too little, and what constitutes too much or too little is determined by the quality of the land. For example, lupin should be cut on gravelly soil when it is in its second flower and turned in while it is still tender. In sticky soils, it should be cut in the third flower so that it provides more organic matter to break up the clods of earth. Pliny quotes Columella as to the value of bird droppings and the dung of cattle: "However, it is universally agreed that no manure is more beneficial than a crop of lupin turned in by the plough" and ". . . where there are no cattle they believe in using the stubble itself or even bracken for manure."

All these topics are covered by Columella just halfway through the second of ten books. The remaining books cover all aspects of plant and animal (including bee) culture and do not concern us here.

Conclusions

What is evident is that the Romans, like all the ancient settled civilisations, were obliged to take steps to maintain soil fertility. Sometimes, this was implicit in the agricultural systems they adopted, such as when the land was regularly flooded, but this was not always the case. Records show that famine was a recurring problem in China as in Rome, and possibly it had a major influence in destroying Classical Maya civilisation. Probably, the populations expanded to consume all the available food supplies, after which there was famine, contraction of the population, and retrenchment. It is not obvious that agricultural productivity necessarily increased in parallel with population, so populations probably reached a constant optimal maximum, overshot the optimum, and then declined before growing again, in a continuous sequence, like a sine wave. Of all the nutrients that plants and animals require, nitrogen is one of the most important. Certainly, in modern, mechanised Western agriculture it is the supply of nitrogen that limits productivity, and it is probable that this has always been the case whatever agricultural system has been developed. It may be argued that the supply of fixed nitrogen has determined the level of population and the quality of life at whatever level throughout history.

What also seems to be the case is that people everywhere appear to have been aware that the productivity of their soils was limited, if only by water, and that where water was not limiting, then other factors were operating. How this was explained is generally not known, but the measures taken to overcome these productivity limitations were similar in the Roman and Chinese empires, and, as we shall see, in Western Europe and in South America.

Famine did not disappear from Europe with the Romans, despite their very considered approach to farming. It recurred frequently, both before and after their epoch. Famine has been a threat to mankind throughout history. Agricultural productivity did not increase beyond a fairly basic subsistence level until farming became more scientific. This was only possible as understanding of the basic requirements of plant and animal growth were discovered. In the meantime, as with the Reverend Malthus and even the Club of Rome, the Cassandras assembled through the centuries to warn of impending doom. Population was always about to overtake the maximum possible food supply. The forebodings have never been fulfilled, at least not in the way that was envisaged. Today there is still no unavoidable problem to feeding adequately the population of the world. How this came about is discussed in the following chapters.

CHAPTER 3

The Development of English Agriculture and the Recognition of the Fertiliser Deficit

Background

So far, it has been shown that nitrogen is an important constituent of living things and that its availability may be limited if soil is not carefully husbanded. Older civilisations have discovered ways of supplying the required nitrogen through the use of fallows or fertilisers of different kinds, such as green fertilisers or manures. Such knowledge must have developed only slowly. The Romans occupied a vast swathe of Europe and parts of Asia and applied their agricultural techniques, as described in the previous chapter, wherever they found them useful. They had realised that a two-crop rotation was not very productive in that half the cultivable fields were lying fallow at any one time. They developed three-field rotations and applied fertilisers and green manures. However, they never succeeded in avoiding the presence of fallow altogether.[1]

This chapter concentrates primarily on the British Isles, for which a vast amount of information is available. Historians of agriculture tend to believe that even Bronze Age farmers were aware of the value of dung and that they probably practised primitive two-field rotations. They did not have a plough other than the ard, which was essentially a hard-pointed shaft that was dragged through the soil by oxen or even pushed through the soil by human brute force. More sophisticated ploughs did not reach Britain until later. Traces of the ancient fields or terraces are still evident in parts of Scotland and on the chalk hills of southern England. These are in open areas that must have been the most appropriate for agriculture since much of the rest of lowland Britain was covered by dense forest, as in the Weald of Sussex and Kent. The Weald (the name derives from the same root as the modern German word *Wald*, meaning wood or forest) is the area of the two counties between the chalk hills of the North and South Downs. These hills are seemingly contrarily called Downs because the name is derived from a Germanic word for dunes. They have been settled for many thousands of years, and the remains of hill forts are evident in many places (figure 3.1).[2]

The fields that were worked in the Bronze Age—that is, after about 1000 B.C.—can still be identified at various places on the Weald and the Downs, and an example is shown in figure 3.2. These fields developed into terraces as they were ploughed year upon year, and their areas represent what could be worked by a single plough team working with an ard in a day, about one acre. They have generally been so eroded that they can be seen clearly only when the sun is very low. Only with the Romans did a plough with a board to turn the soil over, and possibly even with wheels, arrive in England.

The Romans in Britain

The Romans adapted their agricultural systems to Britain, and they were clearly more productive than those of the natives had been, though even they could not raise exotic fruit, such as figs, in Britain. However, they certainly

Figure 3.1. A hill fort on the Downs in southern England, Maiden Castle. This is an enormous construction, the largest Iron Age hill fort in Europe, with ramparts 20 metres high and covering an area of about 20 hectares. Construction seems to have started about 5000 years ago. Reproduced with permission of the Dorset Natural History and Archaeological Society.

Figure 3.2. The remains of cultivation terraces, perhaps as old as 3000 years (Bronze Age) on the South Downs of Sussex near Brighton, England. The terraces can be seen as almost horizontal raised areas, perhaps six in all, arranged down the slope and much eroded by time and weather. The sun is low and to the left, and the shadows across the picture are very long. Photo by the author.

grew grapes and brewed wine, and produce, in particular wheat, was exported from England to the imperial provinces on the continental mainland. Whether this export was a true surplus to local requirements or simply what the estate owners wanted to sell at a profit is not clear. The original Britons had already discovered the crops, such as wheat and barley, that could be grown successfully in Britain.[1, 2]

It is unlikely that the Roman system of agriculture collapsed immediately when the Roman army of occupation left Britain after A.D. 410. There were many Romano-British farmers who had no option but to remain and to carry on with their lives, but the ordered Roman system clearly went into decay. What happened to agriculture in this period is not very evident, but the popular picture of the Dark Ages seems to imply that the inhabitants continued to scratch a living from fields but that they did not serve a highly ordered economic system as was present during the Roman occupation. After the Romans left, villages aimed only at self-sufficiency, and opportunities for commerce with the few small towns were infrequent.

Agriculture in Britain after the Romans

What historians have surmised and gleaned from such records as are extant from this early period is that there was a relatively good supply of cultivatable land in many villages and that this land was often farmed on an open-field strip system. The head of a household, in common with all the other household heads, was allocated strips of land, probably with an area of one acre (about one-half of a hectare), scattered throughout the area exploited by the village. The strips were often of regular shape and size and separated by raised balks or ridges. A typical strip might be a furlong ("furrow long," 220 yards, or about 200 metres) long and a chain (22 yards, four rods, or about 20 metres) wide; that is, of an area of one acre. These dimensions are the basis of the British Imperial system of length. The strips were generally worked in common and in some places were reassigned every season. Probably, some further land was permanently used for pasture. When the crop productivity dropped, it was possible to abandon the land and move on to fresh areas. This kind of land use was possible as long as the population pressure was not great. There is no record of manuring techniques or the exploitation of the nitrogen-fixing properties of legumes, though the value of manure was apparently recognised, and, in order to fertilise them, fields were opened up to sheep and cattle after the crop had been harvested.

Despite the frequent occurrence of famine, the population did grow. According to Prothero[2] in *English Farming Past and Present*: "There was little to mitigate, either for men or beasts, the horrors of winter scarcity . . . On land which was inadequately manured, and on which neither field-turnips nor clovers were known till centuries later, there could be no middle course between the exhaustion of continuous cropping and the rest-cure of barrenness . . . Famine trod on the heels of feasting." Plague was also a frequent visitor. The impression of many people that the Black Death was an unique occurrence is quite misleading. Perhaps the plagues were not all bubonic plague like the Black Death, supposedly brought to Western Europe from Asia by the black rat and its associated fleas, but fatal epidemics of one kind or another were common. The records suggest that from about A.D. 300 until the sixteenth century there were at least five periods of famine every century. William Langland, in the *Vision of William concerning Piers Plowman*,[3] a poem written in the middle of the fourteenth century, quite late in this period, apparently describes a typical situation of a smallholder of that time:

> I have no penny, quoth Piers, pullets for to buy
> Neither geese nor pigs, but two green cheeses
> A few curds and cream and an oaten cake
> And two loaves of beans and bran to bake . . .
> . . . no salt bacon . . . no egg . . .
> But I have parsley and leeks, and many cabbages,

And eke a cow and a calf and a cart-mare
To draw afield my dung, the while the drought lasteth
And by this livelihood we must live till Lammas-tide*
And by that I hope to have harvest in my croft

The consequence of the growth of population was that the supply of cultivatable land became exhausted and more efficient patterns of land use had to be developed. Farmers abandoned the early two-field rotation, in which one field was left fallow in order to restore its fertility whilst a second was cultivated using the normal strip technique, for a three-field rotation. This implies a more sophisticated appreciation of the problem of soil fertility. In such a system, a field with spring-sown wheat or beans might rotate with another sown with autumn-sown wheat and a third fallow field. Although four-field rotations were not unknown, they seem to have been rare. In this way, the characteristic pattern of English mediaeval open-field agriculture developed.

The Normans and After

The organisation of the country and its resources becomes much more evident after the Normans conquered England in 1066. At the instigation of William (the Conqueror), in 1086 they made their extraordinary census of the whole country, *The Domesday Book*.[4] From this survey, one can gain a complete picture of the agricultural resources of the country. It was a very different organisation from that used by the Romans. England was divided into various small administrative units, but the basic agricultural entity was the manor. The manor was ruled over by a lord, who lived in the manor house. There would have been an associated village together with its church and mill for grinding corn. The manorial lands surrounded the village. The inhabitants of the village were not freemen. Most were serfs or villeins, which meant that they owed a feudal allegiance to the lord of the manor. They had to work the lord's land for a certain number of days in the year, and for the rest of the time they could work land that belonged to the lord but that was allotted for their own support (figure 3.3). There was another group called bordars, who had smaller plots for their own use but correspondingly smaller obligations to the lord. Finally, there were some slaves and also some freemen. In England within the Norman domains in 1086, there were some 283,242 male persons. Women and children were not counted.[2, 4]

The ideal mediaeval state seems to have had the Church, sanctioned by God, at its pinnacle, immediately over the king (though some kings seem to have disputed this ordering from time to time), and the king was supported by his barons and lords. The lords owned the manors, and the income that

*Lammas-tide was August 1.

they generated from their serfs supported the whole structure. The serfs paid for their use of the fields by labour on the lord's land. The lord paid for his use of the land by supplying soldiers and hospitality to the king, as and when demanded. The soldiers were often none other than the serfs. Since every person in this pyramid had obligations, both to those above and those below, the interests of all members of society were protected. In theory, the Crown was the ultimate owner of all the land. In England, the arrangement has been called the manorial system. Clearly, some people found this setup more comfortable than others, and it is unlikely that the villeins or serfs operating at the lowest level found it very rewarding.

Large, open fields divided into strips were characteristic of this kind of farming, and it has survived until the present day in at least one part of England. Traces of it can be seen in many places. The "tofts" indicated in figure 3.3 were

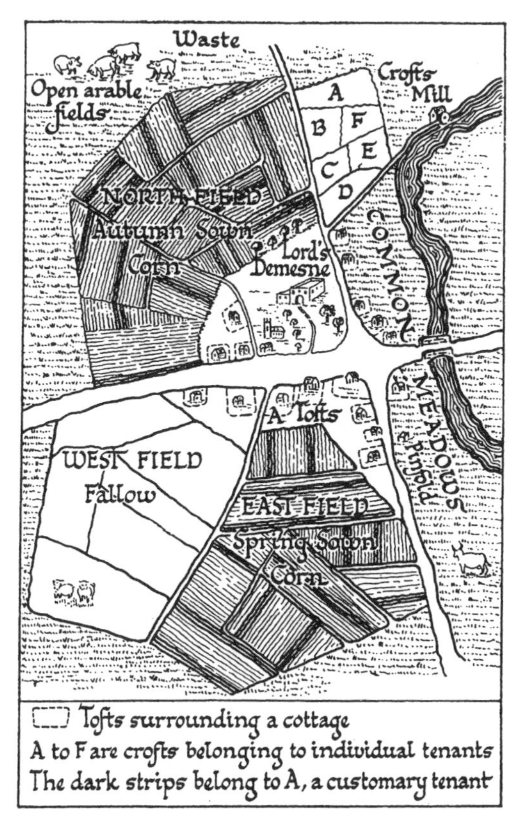

Figure 3.3. A typical open-field village three-field arrangement. The manor was at the centre of the village and is shown here beside the church. The rest of the area served to support these two institutions. Tofts are private plots, and one of the three fields is shown as fallow, probably used for grazing. Taken from J. Orr, *A Short History of British Agriculture*, Oxford University Press, London, 1922.

small, enclosed areas associated with individual dwellings. As far as we can judge, yields from this kind of agriculture were relatively low, and hunger was a constant menace. In some European countries, rather more complex rotations with fewer fallow periods were practised. It seems to have been widely recognised that the ploughing of fallows was useful, if only to destroy weeds, and that manure and composts helped soil productivity.

Mediaeval open-field strip farming has survived until the present day in at least one place in England, at Laxton in Nottinghamshire, centuries after the manorial system and the castle that enforced it have disappeared.[5] There are at present still fourteen farmers who cultivate three open fields in common (figure 3.4), the tenants holding their land in strips. The total area farmed is about 480 acres (about 200 hectares), and this method of exploitation, which originated in the eleventh and twelfth centuries, is now protected and is likely to continue indefinitely.

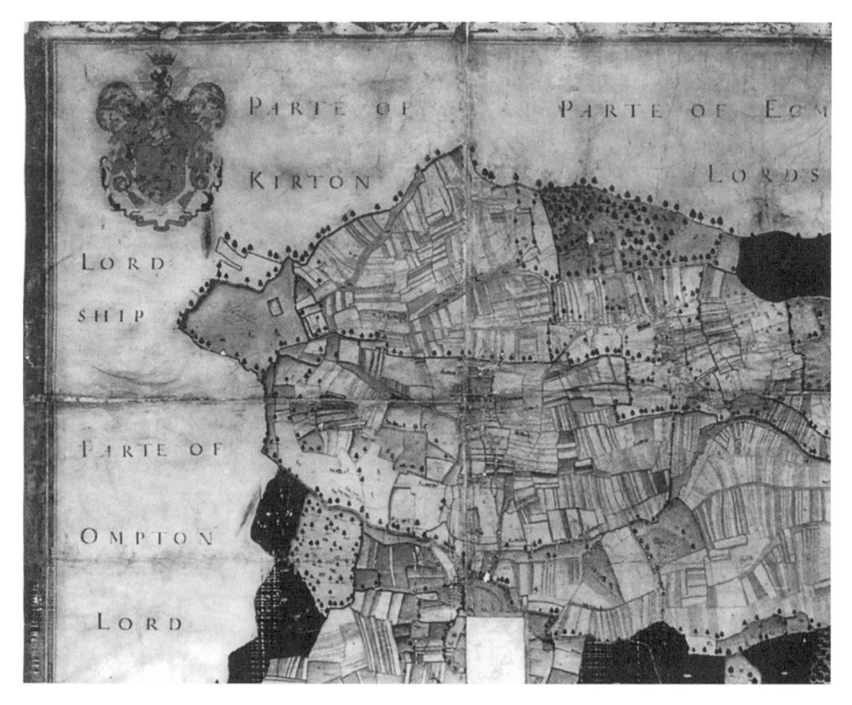

Figure 3.4. A part of a map of the field system around the village of Laxton, Nottinghamshire, England, drawn by Mark Pierce in 1635. The individual strips are clearly delineated. The system has survived here until the present day and is currently worked by 14 families. The original map is now in the Bodleian Library, University of Oxford (Ms. Maps Notts a. 2), with whose permission this is reproduced.

The Influence of the Black Death

The Black Death reached England in 1347, and it precipitated the collapse of the manorial system. The population in England dropped by as much as 50% on average, perhaps most severely in the towns, and one often-presumed consequence, the removal of side aisles from many mediaeval English country churches because the accommodation they provided was no longer required, can be seen to this day. However, this removal has also been ascribed to changes in church liturgy, so the former explanation may not be valid. Nevertheless, some villages simply ceased to exist. In other countries of Europe the overall impact was even more severe.

One immediate consequence of the Black Death was that there were not enough surviving serfs to maintain the lords of the manors in the style to which they were accustomed. This was probably the ultimate cause of the breakup of the manorial system and the development of the class of the landed gentry of Britain, who ran their estates to make a profit rather than simply to provide enough food for the dependent population. In a sense, this represents a change from the Chinese attitude toward the land as the basis of a stable society back to the Roman attitude in which land was a possession to be enjoyed and to be exploited for profit.

In fact, rather than a sudden and universal collapse of the manorial system, there seems to have been a steady development in which some of the peasants enclosed their plots of land, and in which the lords did likewise. A lord would lease the land to his tenants for cash rather than for labour, and he employed landless peasants to work for him, again for cash. The lord could raise some cash by selling his admittedly small surplus to the local towns. This system seems to have been more efficient and easier to operate than the original system of obligatory labour. The political and social consequences of the Black Death are often seen as changing the course of English history, but it probably only accelerated changes that were already occurring slowly. With not enough hands to work the land, much of it reverted directly to the exploitation of the lords of the manors. However, the land was not of much value without serfs, so there was a great pressure to employ the few freemen to work the land for cash payment. At the same time, there was a considerable temptation for the serfs to try to escape from bondage on their home manors, either to work as labourers in the growing towns or as freemen working on other distant manors. In either case, they worked for money wages, and since labour was scarce, these labourers were in a strong bargaining position. They were able to increase the price of their labour, and this threatened the stability of the pyramidal social system. It is ironic that this kind of situation, in which those who are comfortable struggle to prevent change and those who wish to better themselves are forced to rebel, recurs time and again throughout history. As always, progress is acceptable to many people, but only as long as it does not upset their privileges and the status quo!

As early as 1349 King Edward III issued the Ordinances of Labourers that required all people who were not living on their own resources to work the land, and at a rate of pay current in 1346, that is, before the Black Death struck. This did not solve the labour shortage, and it did not satisfy the paid labourers. A series of similar proclamations and Acts of Parliament, sometimes violently enforced, attempted to stabilise the situation. With falling rents and the rising cost of labour and of food, the authorities acted and the peasants reacted[6] and 1381 saw the Peasants' Revolt. This has sometimes been romanticised, however unlikely this may be, by the political Left as a proto-socialist revolt. For example, the story of one leader, John Ball (executed in 1382), an itinerant priest to whom is attributed the couplet "When Adam delved and Eve span, who was then the gentleman?", was retold centuries later in a poem by William Morris.[7] The rebels sacked London, but their leader, Wat Tyler, was killed, and the achievements of the rebellion were nugatory. The landowners continued to consolidate their land in larger and larger enclosed holdings that were easier to work, and they employed more paid labour to do so. The land they could not work was leased out. There is little evidence that agricultural practice improved. That could only happen when the landowners themselves generally became interested in agricultural methods and techniques, and this seems to have been evident only in the seventeenth century.

One great difference between the agricultures of the Roman and other periods discussed in chapter 2 and English agriculture of this period was that livestock was as important for the English landowner as were grain and vegetables. The wealth of Elizabethan England was founded on wool (and, it must be admitted, to some extent on piracy). This meant that sheep farming became a prime interest. The Englishman, at least of the upper classes, ate a lot of red meat, and was proud of it. This must have meant that there was a lot of potential fertiliser available. Keeping the animals through the winter was a problem since there was no guarantee that fodder would last, but that was a problem for the common folk, too. More land was enclosed, and what could not be usefully cultivated was often used as pasture. By about 1485, the feudal system of land tenure had virtually disappeared and the modern system of landholding and exploitation was beginning to take shape. The relationship between landowner and labourer was not one of mutual obligation but was now expressed in terms of cash, paid as wages or rent, which was then exchanged for food. The enclosures of land were clearly not beneficial to all the population all the time, and there appears to have been unrest and even attempts by Parliament to stem the tide. Whereas enclosure was often regarded as inimical to many members of society in the fifteenth and sixteenth centuries, by the eighteenth century it was often trumpeted as a necessity to increase agricultural production, to improve the land. The aim of the new landed gentry was to produce, for profit, grain and meat for the towns and their growing labouring populations. In a sense, one might argue that without land enclosures and the freeing of people to work in the factories, the industrial revolution in Britain would not have been possible. Whether this is true or not, the growth of a class of wealthy landowners with the time and money to investigate how to increase the quality of their livestock and the productivity of their soil, coupled with a more scientific analysis of farming practices, seems to have been the key to rapid agricultural development and to the eventual uncovering of the value of nitrogen for living things and the existence of biological nitrogen fixation.[2]

Estate Management, English Style

As the landed gentry became more numerous, they, like their Roman predecessors, were in need of instruction in how to manage their estates. To fulfill this need, more enterprising farmers produced books of instruction. The verbal tradition of passing down information from peasant generation to peasant generation was no longer sufficient. One of the earliest treatises was written at the end of the twelfth century or the beginning of the thirteenth by a shadowy figure called Walter of Henley, who may have been a bailiff. He seems in some ways to anticipate the changes that were to occur with the breakdown of the feudal system. Walter of Henley describes an essentially feudal estate operating both two-year and three-year rotations and employs the didactic style so popular with Roman writers.[8] He advises on how to select labourers, rather in the Roman style. He has a small section on manure, so clearly its value was recognised even then: "Do not sell your stubble . . . or . . . you will lose much. Good son, cause manure to be gathered in heaps and mixed with earth . . . And before the drought of March comes let your manure, which has been scattered within the court and without, be gathered together . . . Put your manure which has been mixed with earth on sandy ground . . . Manure your lands and do not plough them too deeply." Walter seems to ascribe the benefit of manuring to its ability to retain moisture in dry weather, especially if it is mixed with marl or soil.

A rather more interesting (and amusing) writer is Thomas Tusser, who lived maybe 300 years later, from about 1524 to 1580. He published *Five Hundred Points of Good Husbandry* in 1557,[9] when the manorial system with open fields had passed away. His contribution was written in verse, doggerel of a not very elevated kind. Clearly, he was writing for people who regarded themselves as landowners able to give orders directly to subordinates rather than for manorial lords ("Let pasture be stored and fenced about"), though the idea that these gentry wandered around their estates clutching books of poetry for instruction seems rather far-fetched. The poems are arranged month by month. Somewhat unexpectedly, the agricultural year seems to start in September in Tusser's presentation. He is full of good advice. In January, you must plan your crops.

> In winter a fallow some love to bestow
> Where pease for thy pot they intend for to sow.

By April, you are ready to work.

> The land is well harted with helpe of the fold
> For one or two crops, if so long it will hold

The phrase "with helpe of the fold" seems to imply that animals were allowed to feed from the fallow land and that this helped soil fertility. In May, things begin to get serious.

> In Maie is good sowing, thy buck or thy branke
> That black is as pepper, and smelleth so ranke,
> It is to thy land, as a comfort or muck
> And al thing it is as fat as a buck.

Quite what that means is not exactly clear now because some of the words are obsolete. Buck and branke could be plants, the first perhaps buckwheat.

> Trifallow once ended, get tumbrel and man
> And compas that fallow as soon as you can.
> Let skilfull bestow it where need is upon
> More profit to follow the sooner upon.

A bit later, he writes: "Hide hedlands with muck, if ye will to the knees . . . compas it then, is a husbandlie part." Compas meant manure, as verb or noun. Trifallow meant ploughing the fallow, in this case three times, simply to destroy the weeds. In July, you must "Thryfallow betime, for destroieng of weede."

August was the time to reactivate the fallow, apparently with winter wheat or barley.

> Thryfallowing once ended, go strike by and by,
> Both wheat land and barlie, and so let it lie
> And as ye have leisure, go compas the same
> When up ye do lay it, more fruitful to frame.

In September, you put up fences, and then October is for preparing the new fallow. You must be careful.

> Who laieth up fallow too soone or too wet,
> With noiances many doth barley beset.
> For weede and water so soketh and sucks
> That goodness from either it utterly plucks.

However, it still pays to lay up areas as fallows because:

> Still crop upon crop many fermers do take
> And reape little profit for greedines sake.
> Though bread corne and drink corne such croppers do stand
> Count pease or brank as a comfort to land.
> White wheat upon pease doth grow as he wold
> But fallow is best if we did as we shold.

These lines, and others from the same document, do seem to suggest that the legume pease, a variety of pea, increases the productivity of a field following planting of wheat, though whether this is recommended as best practice is not clear. Corn (in the European sense rather than the American) was used both for bread and for brewing beer, which was a staple drink since most water was generally too contaminated to be safely drunk. Although December seems to be given over to festivities, November necessitates some less pleasant duties that should be carried out when no one can see you at work.

> Fowle privies are now to be cleansed and fide.
> Let night be appointed such baggage to hide
> Which buried in gardens, in trenches alowe,
> Shall make many things better to growe.

And while your hands and clothes are still dirty ". . . Lay compas up handsomely, round on a hill." In other words, build a compost heap.

A further writer of about the same period is another shadowy person known as Master Fitzherbert. He seems to have been alive in about 1520, but, like Walter of Henley, he cannot be identified precisely. He was apparently a landed gentleman of some sort, and his advice covers most of the practicalities of farming, as well as other subjects that would not have been out of place in a Roman account.[10] However, he does not try to seduce his readers with poetry. He uses titles such as: "A shorte information for a yonge gentyl-man. That entendeth to thryue" and "A prologue for the wyues occupation." Note that the letters "u" and "v" were to some degree interchangeable.

The book was originally published in 1534. In it, Master Fitzherbert tells the reader that he is drawing on personal experience of some forty years of farming. He instructs the estate holder how to arrange a fallow: "Nowe these housbandes haue sowen theyr pees, beanes, barley, and otes, and harrowed them, it is the best tyme to falowe, in the later ende of Marche and Apryll, for whete, rye, and barley. And lette the husbande do the beste he can, to plowe a brode forowe and a depe, soo that he turne it cleane, and lay it flat, that it rere not on the edge: the whiche shall destroy all the thistils and wedes . . . for the plough goth vndernethe the rootes of all maner of wedes, and tourneth the roote vpwarde, that it maye not grow."

Master Fitzherbert is against fallowing through the winter because the rain will wash all the "dounge" away and compact the soil, as well as allowing weeds to establish themselves before the wheat is finally sown, all of which would still be regarded as good advice today. He also tells the "husbande" how to "cary oute donge or muck and to sprede it." He advises him to spread the dung "on his barley-grounde . . . at the later ende of Apryll, and at the begynnynge of Maye" and to do this every "seconde falowe," but after the soil has been "sturred," which the *Oxford English Dictionary* suggests to mean ploughing across the former furrows. In this case, it clearly means to flatten the surface and remove depressions. Master Fitzherbert finishes this section with an unacknowledged allusion to older writers. "Horse-donge is the worste donge that is. The donge of all maner of catell, that chewe theyr cudde is verye good. And the dounge of douues [*sic* presumably doves] is the best, but it must be layde vppon the ground very thynne."

What is clearly evident is that manure was regarded as indispensable for good cropping. When the first English settlements in America were planned, it was evident that farming specialists would be required to produce good harvests in soils that had never been manured. In 1585, Sir Richard Grenville reported to Sir Walter Raleigh that the settlement at Roanoke had been planted with "such cattell and beasts as are fitte and necessary for manuringe the countrey."[11]

The Beginnings of Scientific Agriculture

The seventeenth century saw the development of more sophisticated rotations, for the first time in England involving turnips, and legumes such as clover, apparently introduced from Flanders. This was ascribed[12] to Sir Richard Weston, late of Sutton in the county of Surrey, whose work was collected and published by Samuel Hartlib in 1651 in a book called *Legacy of Husbandry*.[13] Weston had visited Flanders and saw that the farmers there housed their cattle in winter and summer and fed them on rape, linseed, and turnips. What amazed him was that the soil of these areas, rather poor and "heathie," was as productive as any he had seen in much richer areas at home in England. He wrote an account of his travels as a "legacy to his sonnes" in 1645, and Hartlib seems to have adopted it as part of a tract to encourage the improvement of husbandry and industry in England. Weston described a complex five-year rotation suitable for use on an English estate that involved legumes such as sainfoin, trefoil, clover, and lucerne, and also intercropping, for example, sainfoin with oats. The use of sheep to fertilise the soil with their dung was also described. In England, the importance of turnips, imported by Weston from Flanders, is enormous because they provided a source of winter food for humans, as well as for the livestock, particularly for sheep. It was no longer necessary to kill all but the breeding stock as winter set in. Large flocks could guarantee a reliable supply of wool, milk, and meat. They also contributed to the wealth of the realm.

Hartlib's version even carries an advertisement: "If any desire to have the great Clover of Flaunders, or the best sorts of Hemp and Flax—seeds of those parts, or Saint-Foine, La Lucerne, Canary seeds or any Seeds of this kinde: let them enquire of Mr. James Long's Shop at the Barge on Billingsgate." Billingsgate is still a defined area of London.

Weston notes that some of his recommended practices were already being adopted in some parts of England. The use of the new crops and methods slowly increased, and they allowed the development of the four-course, or Norfolk, rotation, generally ascribed to the second Viscount Townshend (figure 3.5). Though he was not the first to advocate the use of turnips, his name

Figure 3.5. Charles, second Viscount Townshend (1675–1738), from a portrait by Godfrey Kneller. The Viscount was a statesman and an agriculturalist. On account of the latter, he was also known as Turnip Townshend. He is considered responsible for introducing the Norfolk four-crop rotation into Britain and for emphasising the value of turnips as a feed for animals and for humans, hence his nickname (portrait reproduced by kind permission of Lord Townshend).

became associated with their use, and he was often referred to as "Turnip Townshend." In the four-course rotation, cropped fallow replaces bare fallow, and clover and grasses were introduced after spring-sown corn. The clover meant more hay and grazing, the turnip meant more winter feed and more livestock, and that, in turn, meant more meat and more manure.

These changes were probably taking place all over Britain at different times and in different places. A remarkable record has been left by one Bulstrode Whitelock,[14] who was a Member of Parliament and even acting President of the Council of State of the Commonwealth of England in 1651, 1652, and again in 1659. His diary, *Memorials of the English Affairs from the Beginning of the Reign of Charles the First to the Happy Restoration of King Charles the Second* (roughly from 1625 until 1660), is essentially a summary of letters to Parliament during that period. Many throw an interesting light on political affairs of the time. For example, on April 4, 1650, there was a letter "from the diggers and planters of commons, for universal freedom, to make the earth a common treasury, that every one may enjoy food and raiment freely by his labour upon the earth without paying rents or homage to any fellow-creature of his own kind." What the landed gentlemen of Parliament made of this appeal of proto-communists is not recorded. Cromwell's New Model Army invaded Scotland in 1650 and, in a reflection on the local agriculture, reported from East Lothian on September 6 that "in those parts where the army marched was the greatest plenty of corn [wheat] that they ever saw, and not one fallow field, and now extremely trodden down and wasted, and the soldiers enforced to give the wheat to their horses." Elsewhere they mention that they lacked oats and hay to feed their horses, so this must have been an unusual thing to use. Notwithstanding, apparently the Scots of East Lothian knew how to increase productivity and to avoid the use of fallow. How this was achieved is not clear, but maybe they had learned from Flanders, as others were to do later. The close connections of Scotland with the continental mainland had made it a country in many ways more sophisticated at that time than England. Some would claim that it still is.

The period after about 1650 saw the beginning of the construction of the foundation of modern science and, in this particular context, of chemistry. Hitherto people had taken the basis of all matter to be the four elements earth, air, fire, and water. The most eminent figures of the period, such as Newton and Boyle, were interested in alchemy, which is not to say that such facts give any credence to that subject any more than the fact that almost everyone at that time believed in astrology gives any modern credence to it. The problem of finding a rational way to understand what exactly a fertiliser or dung actually did is well-illustrated by an experiment reported by an eminent Dutch scientist, J. B. van Helmont (figure 3.6). This was at a time when chemists were beginning to realise the virtue of carrying out quantitative experiments of the kind later used so beautifully by the French chemist Antoine-Laurent Lavoisier.

Van Helmont's famous tree experiment[15] was widely quoted to confirm that the element water could be used to make the matter of a tree, roots, leaves, and branches. The experiment was reported in an interesting newspaper pub-

Figure 3.6. J. B. van Helmont (1579–1644), chemist, medical man, and alchemist. His son probably had two eyes, but the unfortunate juxtaposition in this drawing of the two heads makes this unsure! The original is contained in his book, *Ortus Medicinae*, published in Amsterdam in 1652. This version was taken from J. Read, *Humour and Humanism in Chemistry*, G. Bell and Sons, London, 1947, facing p. 70.

lished by one John Houghton, F.R.S., at irregular intervals over about ten years, from 1692 until about 1703.[16] This was called *A Collection for Improvement of Husbandry and Trade*, published in weekly instalments (figure 3.7), and its intention is clear from the title. Its contents were quite eclectic, dealing with matters ranging from the size of Earth to the number of houses in various cities of England, the price of meat, and the constituents of urine. In particular, he intended to describe "the best sort of Compost or Manure for each (sort of Earth); and how to be cultivated". As for van Helmont, on Wednesday, May 4, 1692, in issue number 8, Houghton wrote that: "He took two Hundred pound weight of Earth dryed in an Oven, and put it into a Vessel, in which he let a Willow-Tree which weighed five Pounds, which, by the addition of Water to the Earth, did, in five Years time, grow to such a bigness, as that it weighed one Hundred and sixty nine Pounds; at which time also he dryed and weighed the Earth, and within two Ounces it retained its former Weight. This I had from French's Art of Distillation, pag. 130." Houghton draws the conclusion that Earth, being gross, could certainly not enter the "small holes (Pores)" of the roots of tender plants and thus could not have contributed to the nourishment of the tree. The unstated conclusion, which was also drawn by van Helmont, was that the water had become transmuted into the material of the tree.

The experiment of van Helmont was a very good one, and the conclusion drawn, in the light of contemporary knowledge, was eminently reasonable. It is also, but only in hindsight, incorrect. Such misinterpretation was inevitable, given the ideas that were then extant. For example, van Helmont was also an alchemist,[15] and he had once apparently received a sample of the philosophers' stone from a stranger, perhaps the famous Scottish alchemist, Alexander Seton, whom he had met by chance. Seton was apparently one of the most successful alchemists, and van Helmont was acknowledged as an eminent scientist. Van Helmont reported: "I have at distinct turns made projection with my hand, of one grain of the Powder, upon some thousand grains of hot Quick-silver." Further, the philosophers' stone is "Saffron in its powder, yet weighty, and shining like unto powdered Glass: There was once given unto me one fourth part of one Grain: But I call a Grain the six hundredth part on one Ounce." In 1618, and apparently in the sight of several onlookers, he transformed eight ounces of mercury into eight ounces and eleven grains of purest gold. This implies a ratio of gold to philosophers' stone of 19,186.

It is not clear what one is to make of reports like this. Similar friendly strangers seem to have met other people besides van Helmont, and the stone is never properly characterised. However, van Helmont was certainly able to test gold for purity in a reliable manner, and he appears to have carried out the transmutations in the absence of his friendly stranger and in the presence of witnesses. No wonder that the transmutation of water into the material of a tree was no great news, and no wonder that Newton and Boyle were not sceptical of such reports.

Vol. 6. Numb. 129

A

COLLECTION

For Improvement of

Husbandry and Trade.

Friday, January 18. 169$\frac{7}{8}$.

How Skins *are prepared to beat* Gold *in*. Blood *good for* Swine, *and nouriſhing*. *It purifies* Salt. Whites *of* Eggs *fines divers things*, *and* Ivory-Wort. Blood *is excellent Dung*, *particularly for* Fruit-Trees *and* Wheat, *as is ſhewn by an Inſtance taken from Mr.* Evelin. *Price of* Corn, &c. *in many Counties*. *Of* Engliſh Commodities *in* London. *Of* Actions *in* Companies. *Of* Fleſh, &c. Schools, Publick Notaries *and* Woodmongers. Exchequer Funds. Advertiſements.

IN Order to prepare the Leaves for *Gold-beaters*, ſome Women pull off the Skins of the *Rectum*, (*Streight-Gut*) and ſell them to the *Gold beaters* for ſix Shillings the hundred. They are from one to three Foot long, and about ten Inches wide, and about as thick as a blown Bladder ; 'Tis the laſt Gut of all, and that which makes the *Roll-Tripe*, or is of a Hog call'd the *Inch-Pin*.

Theſe the *Gold-beaters* wet in Water, then open them and lay them on an Engine call'd a *Ladder*, without Barrs, except one in the middle, only a long Frame of ſquare Pieces. On this they lay two together, joyning on the ſmooth Sides; then they dry them, and when ſo, the Greaſe and rough Matter is taken off with a *Pumice-Stone*, and after that rub'd with a Spunge dip'd in a Mixture made with *Aqua-vitæ*, the fineſt *Gum-Arabeck*, and ſome whole *Cloves*, and it ſtands three or four Days till the *Gum* is diſſolved, and it is as thick as uſually *Gum Water*. The Skins muſt be rub'd with this all alike, and on both the Out-ſides; and when they are dry, they are cut off into Form for uſe.

Beſide the uſe above, they are reckoned excellent to cure cut Fingers ; but whether it be from the Nature of the Skins, or from the Gum, or rather from the fitneſs of the Skin to wrap about it, and keep off external Injuries, I'll leave to Others to determine, altho' many think the Blood is a Balſam of its ſelf.

The Blood is uſeful to many Purpoſes ; 'tis good Food for Swine ; and I ſee nothing but it may be almoſt as nouriſhing as the Fleſh for altho' Urine, Gall, Sweat, Spittle, Tears, Snot and Ear-wax, as Excrements, are caſt from it before it becomes Fleſh, yet a great Part does become ſo, or at leaſt wiſe fills up thoſe Veſſels, which of late Days the Fleſh is ſaid to be made with ; and we find by experience that Black-Puddings are a wholſome Nouriſhment: And I ſee no natural Reaſon why the *Jews* were prohibited Blood : I believe 'twas mainly to make them abhor Murder, and becauſe they ſhould not be too familiar with their Sacrifices.

At the *Wyches* in *Cheſhire*, &c. the Salt is purified with Blood, I preſume for the ſame reaſon we purifie other Matters with Whites of Eggs : and ſome boyl Ivory Shavings in their Wort : 'Tis glutinous, and therefore the Filth ſticks to it, and ſo is carried upwards.

Blood is excellent Dung becauſe 'tis alkalous (as we ſee by Spirit of Blood) and imbibes the Nitre, as indeed all animal Subſtances will.

The moſt Ingenious Mr. *Evelin* in his *Philoſophical Diſcourſe of Earth*, *Pag.* 319. ſays, 'tis excellent almoſt with any Soil where Fruit is planted, eſpecially Mural (Wall-Fruit) of great advantage to the Grape, poured about the Roots deluted. He tells that after the Battel of *Badnam* in *Devonſhire*, won by the Lord *Hopton*, the Blood of the Slain did ſo fertilize the the Fields where Corn had been ſown a little before, that the Year following produc'd ſo extraordinary a Crop, as moſt of the Wheat Stalks bare from two to fourteen Ears. The Owner by reaſon of its Treading thought to have reſown it, but was diſſwaded, and it happened as above.

In *Pag.* 322, he ſays Lime tempered with Blood, extraordinarily recreates it.

Next *Friday* expect more from

Yours *&c.*

John Houghton, F. R. S.

I Am removed to the *Fleece* at the Corner of *Little-Eaſtcheap* in *Grace-church ſtreet*, and deſire all my Correſpondents to direct their Letters thither.

Figure 3.7. Copy of the first page of an early issue of John Houghton's *A Collection for Improvement of Husbandry and Trade*, published at irregular intervals over several years, from about 1690 until about 1700. This was taken from a facsimile published in 1969 by Gregg International Publishers, Hampshire, England.

An even more surprising topic seems to have caught the attention of Houghton, the newspaper publisher. He was always gleaning information from whatever source, sometimes without being too careful about its reliability. On February 2, 1692/93, he writes: "Some in Essex have their Fallow after Turneps, which feeds their Sheep in Winter, by which means their Turneps are scooped, and so made capable to hold Dews and Rain water, which by corrupting imbibes the Nitre of the Air, and when the Shell breaks, it runs about and fertilizes (makes fruitful). By feeding the Sheep the land is dung'd, as if it had been folded." In the same issue, he describes the use of horse and cow dungs to treat fallow and implies that "pease" are used as part of a rotation. The striking thing in this excerpt from the newspaper is the idea of imbibing nitre from the air, though this statement should not be taken at current face value. What was this nitre in the air of seventeenth century Essex, and how was it imbibed by the residues of the "Turneps"?

In 1692, What Was Meant by the Word "Nitre"?

An examination of the whole run of Houghton's newspapers reveals several related statements concerning nitre. A solid called nitre had been known for some centuries and was used for making gunpowder. It seems unlikely that anyone at that time could have had any inkling of its true nature. The nitre described by Houghton was something different, something that resided in the air.

Houghton states in issue 3 of his newspaper that he will report on Earth, Air, Fire, and Water, though "not as Principles or Elements." He is much exercised by the gross nature of earth, and how it could not nourish plants on that account, but he can explain how lime produced by burning chalk, as it still is today, can fertilise them (in the following excerpts, the words in parentheses are just as shown in the original): "And I do presume Lime to be made by a forceable burning away of all the Moisture concealed in the (Minute) small hollowness of the Chalk, or other Material of which it is made, whereby it becomes apt greedily to drink up the Nitre or Spirit of the Air, the which being (diluted) wash'd out by Rain or otherwise, becomes a Food fit to be suckt up by the Plants, in order to their Growth and Nourishment." Since nitrogen was completely unrecognised at the end of the seventeenth century, let alone could it ever have been regarded as an element, such statements cannot contain descriptions of nitrogen fixation.

Nevertheless, the newspaper is full of interesting matter that sheds light on the life of the times. For example, on January 18, 1694 (or 1695), it carries an advertisement to the effect that: "Walsal the Coffee-Man against Cree-Church in Leaden-Hall-Street, keeps a Library in his Coffee-Room for his Customers to read. He also buys and sells Books." Coffee shops doubling as bookshops is not such a new idea!

Houghton addresses the problem of how this aerial nitre dissolves in water: "I know 'tis by a great many ingenious thought, that Nitre, or some other salts flying in the Air (adheres) sticks to the small bubbles of Water and so fixes them into Hail, Snow or Ice . . . and Northern Countries where there is most of the Ice, Snow, or Hail, should yield great store of Nitre, and by consequence their land should be extreme fruitful, but from our Southern Counties we have our Salt-Petre, and they are also much more fertile." This passage is noteworthy for the distinction he makes between nitre, which is not clearly defined, and saltpetre, which we now know to be potassium or sodium nitrate. It was produced by extracting compost heaps containing dung of all kinds, urine, and rotting organic matter with water, and earlier in the seventeenth century it was apparently used exclusively for making gunpowder for military use. By 1692, it was recognised and used as a fertiliser. There has even been a suggestion that the submarine designed by Cornelius Drebbel and reputedly rowed by twelve oarsmen underwater in the River Thames from Greenwich to Westminster in 1621 used heated saltpetre as a source of oxygen to sustain the enclosed rowers, but this is far from sure.

However, there is clearly a connection between nitre and saltpetre. On May 18, 1692, Houghton described an experiment in which he says: "The reason why I put Wood-Ashes to the Snow-Water was, that it might extract its salt. And that the salt so extracted from the Ashes, might collect the little Parts (particles) of the Nitre to it, as is usually done when (Nitre) Saltpetre is drawn out of the earth dug in Cellars." Ashes were a known source of saltpetre, but attempts to find nitre in the residue from distilled melt-water were apparently unsuccessful.

In a subsequent discussion on October 27, 1693, Houghton talks about growing corn and ventures the suggestion: "For Wheat the Clay is plowed four times, but for Pease and Beans but once. The Reason of plowing I suppose . . . [is] that it may with the Rains and Dews receive the Nitre from the Air, which is the chief thing that I believe makes fruitful . . . Now if it be seldom plowed, there can be little part made fit to receive the Nitre . . . The reason why plowed but once for Pease and Beans it's probable is, because the land was so well prepared the year before for Wheat, and the Nitre is not yet spent." Dung apparently works its goodness by storing nitre from the air, but although the effect of legumes in improving soil fertility is hinted at here, Houghton observes "that every third year is lost in Common Fields," presumably a reference to fallowing in a three-year rotation.

So something beneficial for plant growth seems to pass from the air into dung, and thence into plants, and saltpetre can be extracted from ashes and dung and can also be used as a fertiliser. This connection was remarkably prescient since the nature of nitrogen was not established until almost 100 years later. On May 28, 1692, Houghton reported: "Air is a compound Body made up of light and fine (Atoms) small Parts such, as are the (Effluvia) Steams of allsorts of Matter, as Water, Smoak, Plants, Animals, Nitre, Etc. but the much greater part thereof is a pure and native Substance of its own kind, the necessary Food of Life in all sorts of living Creatures, as likewise of Fire; for without it both Flame and Life are instantly extinguished." So nitre is regarded as some kind of impurity in air, and air itself is not a compound body but still a single element. The discovery of nitrogen in the eighteenth century and its relationship to nitre and saltpetre will be discussed in chapter 4, which deals with the development of chemistry.

The Development of Crop Rotations and Manures

From the material quoted above, it is evident that in the 1690s the three-crop rotation was still common. However, the English landed gentry were not ones to lose one-third of their land to fallow, with the consequent loss of income, and as they bred improved cows, horses, and pigs (to develop those curious cuboid animals with one leg at each corner, so proudly displayed in many drawings and paintings) (figure 3.8) they also developed much more sophisticated rotations.

By early Victorian times, the knowledge of complex rotations was widely distributed. For example, one amongst many others, *The Library of Agricultural and Horticultural Knowledge*, published in 1834 in its third edition,[17] ran to some 600 pages, of which eight describe the rotation of crops. It shows proudly on its title page the somewhat outdated quotation from Francis Bacon that must have been no longer completely valid even in 1834: "The improvement of the ground is the most natural way of obtaining riches." It would certainly not be accepted in industrial countries today, even if it may still be applicable in agrarian societies. This book describes a multitude of manures and their most appropriate application. There is a long discussion of night soil, as well as a description of the practices of the Chinese as reported by British diplomats in China in the 1790s. How this night soil was collected in

Figure 3.8. The basis of the roast beef of Old England. Improvements in cattle, pigs, sheep, and horses were achieved by selective breeding. Taken from J. Orr, *A Short History of British Agriculture*, Oxford University Press, London, 1922.

Britain is not stated, but Mr. Hitchins, then the land surveyor of Brighton, applied four wagonloads to the acre. After night soil, pigeons' dung was supposed to be the best manure.

The rotations, also ascribed to Mr. Hitchins, are grouped according to the type of land to be worked. On stiff clays, Mr. Hitchins recommends a ten-year sequence following a fallow that has been well-manured and cleaned of weeds. The rotation is seeds, wheat, beans, oats, seeds, oats, tares to be ploughed in, wheat, beans. This amounts to one fallow in ten years. On best-quality chalk soils, Hitchins recommends fallow for rape to provide fodder, barley, clover for hay, wheat, barley or oats, turnips or rape, wheat, barley, seeds or tares. It should be noted that tares are vetches, a crop that was grown for fodder and for hay and also has the advantage of being a legume. The biblical use of the word, meaning a weed, seems to have been because the translators of the Wycliff Bible at that time regarded tares as weeds, especially amongst wheat, and the value of specific vetches was recognised only later. It seems that by "seeds" was meant clover or a grass. Thus, there were no fallows of the traditional kind.

These rotations would clearly help to increase food production, but that was not at the back of Mr. Hitchins' mind. He writes that land should always be well-manured and kept clean and that the landowner should require his tenants to observe suitable rotations to keep the land in good heart. It is evident that unless an appropriate system of rotation is adopted, the tenant may not be able to pay his rent. He recommends ". . . a clause first introduced into leases . . . restricting tenants from sewing more than two white straw crops in succession . . . [despite] an alarm . . . instantly raised that no tenant could long afford to pay his rent, . . . is now generally adopted by choice, from a conviction that more profit can be thus realized." Clearly, profit was the major consideration. The poor, and especially those in the towns that were now growing rapidly under the pressure of the Industrial Revolution and the land enclosures, often went hungry even if they did not actually starve. Charles Dickens' Oliver Twist spoke for a large number of deprived folk when he asked for more.

The Use of Artificial Fertilisers

One of the ways in which land was maintained in good heart was the liberal application of fertilisers and manures, but it is evident that even the improved animals developed by the breeders could not produce enough dung to do this. By 1834, when *The Library of Agricultural and Horticultural Knowledge* was published, artificial fertilisers were also used widely. Apart from chalk and marls, and even common salt, sal ammoniac (more properly ammonium chloride) was imported from Egypt, where it was distilled from camels' dung. Saltpetre (potassium nitrate) had been recognised as a good fertiliser by the time Houghton was writing his journal 150 years previously, but then it was more valuable as a component of gunpowder, and supplies were restricted. Unlike sal ammoniac, it was produced within the British Isles from dung rather than imported. At least by 1834 it was also imported from natural deposits in Italy, Africa, India, or Syria. Sodium nitrate was already known at that time to exist in large deposits in South America, but it was not widely available for fertiliser use. Saltpetre is referred to in *The Library of Agricultural and Horticultural Knowledge* as "impure or crude nitre," and the recrystallised material was called sal prunella (presumably because the lumps of product were the size and, perhaps, colour of plums). The crude material was said to be adequate for farming purposes. The *Library* states: "It is universally admitted by those who have used it at the rate of one hundred weight (about 50 kg) per acre, to be productive of the most luxuriant effects, and to retain an advantageous influence upon the soil for at least two years."

The Looming Food Crisis: The Reverend Malthus

All these developments, together with the use of legumes for reviving fertility, were not adequate to reconcile the need for profit, the demand for food, and the requirements for fertiliser. The probable consequences of the situation in Britain were enunciated clearly, though not for the first time, by the Reverend Thomas Malthus (figure 3.9), who published his essay *On Population* in 1798.[18] In fact, Benjamin Franklin and others had already broached the subject of population growth.[19] On several occasions, Franklin, initially an ardent supporter of the connection of the American colonies with the

Figure 3.9. The Reverend Thomas Malthus (1766–1834), taken from J. Bonar, *Malthus and his Work*, 1st edition 1885, 2nd edition 1924, new impression Frank Cass and Co., London, 1966.

"Mother" country, pointed out that the population of the American colonies was growing so fast that it would upset the nature of the relationship between England and her dependent territories. However, he did not seem to regard the population growth as a danger. In 1751, he made some observations on the matter. In America, he said, more people were marrying and were marrying earlier and having children younger than in England. This was going to create a new market for British goods. The reason for this expansion was the availability of land, which gave people the opportunity to earn money more quickly, immigrants being unlikely to remain long as labourers in the cities. He was against the import of slaves because he doubted that slave labour was more economic than that of working freemen. In any case, owning slaves was morally debilitating for the slave owners. He clearly opined that if there is space available, then population will expand to fill it, but he didn't speculate on the nature of that space. What happens when a people outgrows its food supply was apparently evident to him. Much of this sounds liberal, even modern. However, Franklin was a man of his time. He was all in favour of immigration into the American colonies, but: ". . . the Number of purely white People in the World . . . is very small. All Africa is black or tawny. Asia chiefly tawny. America (exclusive of the Newcomers) wholly so. And in Europe, the Spaniards, Italians, French, Russians and Swedes, are generally of what we call a swarthy Complexion; as are the Germans also, the Saxons only excepted, . . . I would wish their Numbers (the Body of White people) were increased . . . why increase the Sons of Africa, by planting them in America where we have so fair an opportunity, by excluding all Blacks and Tawneys, of increasing the lovely White and Red?" So Franklin's opposition to slavery was based upon economic arguments and did not, at least in 1751, stem from any significant moral judgment on the system. The same is probably true of many of Malthus' arguments.

It should be remembered that famine and hunger were constant threats to people then as they still are now in some parts of the world. Some philosophers were optimistic about the possibility of improving the human lot, and these people were very well aware of the dreadful lives forced upon the poor. This was a time when Adam Smith's analysis of capitalist society was becoming popular, and his influence even then, let alone later in the Britain of Mrs. Thatcher and the United States of President Reagan, was considerable. It was a matter of reasonable enquiry as to whether population could grow indefinitely, and Malthus should be regarded in that light rather than as a heartless analyst of other peoples' poverty.

The basic question was whether society could ever be perfected so that want, poverty, and vice could be extinguished. Initially, Malthus was clearly of the opinion that it could not. In his famous essay of 1798,[18] Malthus also tried to reach a "scientific" conclusion. He acknowledges his debt to Adam Smith and to David Hume and starts by saying that the "great question" is whether mankind shall move toward an unconceived improvement or be condemned to a perpetual oscillation between misery and happiness: "I have read some of the speculations on the perfectibility of society with great pleasure (and) I ardently wish for such happy improvements." He bases his arguments on two "postulata." The first is that food is necessary to the existence of man, which seems to be undeniable. The second is that passion between the sexes is necessary and will remain nearly in its present state. If the first postulate is likely to be accepted by almost everyone, the second is not necessarily so. One of the proponents of the happy future of mankind, a Mr. Godwin, whose musings in an earlier publication had prompted Malthus to write his essay, "conjectured that passion between the sexes may in time be extinguished." Malthus comments that although mankind has made much progress from its original state of savagery ". . . towards the extinction of passion between the sexes, no progress whatever has hitherto been made." That still seems to be the case today!

Malthus then goes on to say that population, when unchecked, increases in a geometrical ratio, whereas subsistence increases only in an arithmetical ratio. Consequently, sooner or later, population will outgrow food supply. This is his major argument against the perfectibility of society, and he states it right at the beginning of his essay. The remainder of the essay contains a defence of the propositions of arithmetical and geometrical increase. It is on this basis that he opposed the Poor Laws then current in Britain, in that they kept wages lower and prices higher than they should have been. They also allowed the poor to reproduce faster than economic circumstances would justify. "I feel no doubt that the parish laws of England have contributed to raise the price of provisions and to lower the real price of labour . . . The poor-laws of England may therefore be said to diminish both the power and the will to save among the common people . . . The mass of happiness among the common people cannot but be diminished, when one of the strongest checks to idleness and dissipation is thus removed." Finally, misery, by which we must understand poverty, hunger, and disease, limits population. He extends the argument to North America and speculates that primitive hunting as a basis for population is not very secure and that this naturally controls population growth. He even states: "It is said that the passion between the sexes is less ardent among the North American Indians than among any other race of men." The basis for this statement, which apparently also applies to "the Hottentots near the Cape," is not explained.

Similar arguments concerning population and food supply are not unheard of even today. The last and greatest restatement was that of the Club of Rome in 1972.[20] The argument that population growth, if unchecked, must eventually exceed the resources available to support it is incontrovertible. However, it is not possible to be sure at any given point in history that such a situation has arisen. Malthus was clearly wrong then, but he could not have foreseen the developments of the nineteenth century, just as the Club of Rome was wrong more than 150 years later.

Malthus' essay was reviled in some quarters but was also extensively revised by the author by 1803. The seventh edition was published in 1872 (posthumously, for Malthus died in 1834) and is remarkable for being about three times the length of the original essay and also for coming to precisely the opposite conclusion.[18] Presumably, the reception of the original essay was not as enthusiastic as Malthus had hoped. The new version included a much-extended discussion of the growth of populations in a variety of countries and on the limitations to population. The preface to the second edition states that ". . . in the present work I have so far differed in principle from the former, as to suppose another check to population which does not come under the heading of either vice or misery." Toward the end of the revised essay, this check is stated to be "moral restraint." This is not to say that Malthus regarded "passion between the sexes" as being bad, but he clearly sees that love is somewhat more than the indulgence of animal instincts. It is, indeed, "one of the principal ingredients of human happiness." Nevertheless, if moral restraint can be duly exercised, then population can be controlled at a level supportable by the supply of food and goods, and misery and vice can be obviated. Despite this reversal of view and his optimistic conclusions, Malthus is generally remembered only for his first pronouncements concerning the rates of growth of population and food supply. Unfortunately, his remedy for an apparently undeniable problem was, and probably still is, too idealistic to be effective in the real world.

New Sources of Fertiliser: The Trade in Guano

Populations, and the accompanying "misery and vice," still grew uncontrollably, though the problem was less acute in North America where there were further territories to exploit, even if the original native inhabitants needed to be removed first. Agricultural productivity also rose, but further external sources of fertiliser were necessary to maintain the momentum. With the growth of the European empires and the development of global commerce, new sources of fertiliser became available, particularly in the Americas. The first new importation of fertiliser from outside Europe, and principally into Britain, was from Peru in the form of bird droppings, otherwise known as guano.

Knowledge of the existence of guano was first brought to Europe at the beginning of the seventeenth century. A man called Garcilaso de la Vega, born in Cuzco in 1539, son of an Inca princess and a highborn Spanish Conquistador, came to Spain after 1560 in order to obtain some recompense for the services of his father to the Spanish Crown. Known as El Inca, he published his two-volume *Comentarios Reales* in Lisbon in 1609, though they were apparently written in 1604.[21] The second volume described the conquest of Peru by the Spaniards, but the first was an account of the life of the Incas before the Spaniards came. He described their agriculture, how they constructed terraces and associated irrigation channels on the sides of the mountains, and how land was assigned to the inhabitants. There seems to have been a primitive social security system that ensured that the aged, sick, widows, and orphans were fed, though they did not cultivate the earth.

They "ploughed" using the combined efforts of several persons using pointed sticks forced into the earth by means of a footrest (figure 3.10). They fertilised their fields with manure. Inland they used "human" manure, which was regarded as the best, but near the coast they used the dung of seabirds. Garcilaso describes islands off the coast of Peru (the Chincha Islands) that were composed almost entirely of guano. The Inca rulers allowed their subjects to take quantities of this material to fertilise their cultivated terraces (there were few open fields in the foothills of the Andes), and they specified who should take the guano, how much, and when. It was specifically forbidden, under penalty of death, to remove guano at times when the birds might be disturbed (for example, when they were breeding), and killing the seabirds was also a capital offence.

De la Vega also describes how the Incas in some parts of the country used as fertilisers the heads of the sardines that generally proliferated off the coast of Peru. This coastal area was, and still is, desert, and it rains there only very infrequently. The farmers dug pits that could be of a considerable area down to the water table. They made holes in the bottom of these pits with stakes, and in each hole they burned two or three maize seeds and the heads of sardines. This was sufficient to fertilise the pits and grow seeds. According to de la Vega, these pits did not require watering, nor weeding, nor further manuring. "The

Figure 3.10. The Inca planting corn, as illustrated by Felipe Poma Guaman de Ayala in *El Primer Nueva Corónica y Buen Gobierno*, 1613–1615. This was clearly a tedious process. This book was a bit later than that of de la Vega, and dealt with some of the same material, but was not part of a plea for financial reward. Reproduced with permission from what is apparently the only extant copy of this book, now in the Royal Danish Library, Copenhagen.

Indians are unable to say who invented the system of pits; and the inventor must have been necessity . . . Thus all the natives sowed what they needed for their sustenance and there was no need to sell supplies or to hoard them; and they did not know what want was." Apparently, similar practices were also operated by some North American tribes when the European colonisation began.

The sardines thus fed the people both directly and indirectly. Note that there is sometimes confusion between the names "sardine" and "anchovy," which do not necessarily mean the same kind of fish in different parts of the world, but both kinds are members of the herring tribe. The abundant fishery off the coast of Peru was not recognised in North America, Japan, and Europe as a commercial opportunity until after World War II. A large export industry of fish and fish meal (for use as fertiliser and animal feed) grew up. The catch grew from 92,000 tonnes in 1952 to 10.5 million tonnes in 1961. This rate of growth rapidly exhausted the source that fed it, and the catch has since been restricted. Sadly, this kind of story has been often repeated elsewhere.

It seems that no one in seventeenth century Europe really noticed the implications of de la Vega's account, and it apparently had no resonance. In any case, the Spaniards were more interested in gold than bird dung, even though it was several times stated to be the best of fertilisers. Whether the economics of a trade in guano would have been favourable at that time is also questionable, though the English and Dutch pirates lying in wait for the Spanish treasure ships might not have been pleased to find them full of bird excrement. Nevertheless, one recent commentator wrote that the ornithological guano producers were the world's most valuable birds in their dollar yield for each digestive process!

The great German explorer Alexander von Humboldt brought a sample of guano to Europe in 1804 and had it analysed. It was found to contain very high proportions of nitrogen and phosphorus. Scientific agriculture was only just beginning at this time, but news of this rediscovery of guano by Europeans had a much greater effect than the earlier announcement of de la Vega. This time a trade in guano was established, and although the guano was clearly a good fertiliser, the effects of the trade upon Peru itself were much more significant. Deposits of guano were discovered in other parts of the world, and the exploitation of this valuable source of fixed nitrogen and phosphorus, so ideally placed for shipping to the fields of the industrialised countries, generated great wealth. Unfortunately, the guano had accumulated over millennia, and it was exhausted in less than fifty years, despite the continuing efforts of the seabirds to replenish the stockpiles. The end was obvious and sudden.

In 1820, the Spanish empire in South America was beginning to break up.[22] There were large numbers of persons, and even some genuine patriots among them, who were interested in dividing the spoils of empire. There were many wars between states that had originally been divisions of the old empire, and there was much graft and corruption, often encouraged for their own particular reasons by the United States of America, France, and, particularly by Great Britain. In Peru, there were at least twenty-four changes of regime between 1825 and 1841, but from 1840 to about 1870 there was some stability brought about by the riches from the exploitation of guano and later of nitrate.

Although the beds of guano were hundreds of metres thick, their exploitation was not difficult, and it could be easily loaded and transported as long as it was kept dry. The trade took off in the early 1840s, and the British firm of Antony Gibbs & Sons, with a virtual monopoly on the trade, was paying $15 (U.S.) for a ton and selling it for $50 per ton, primarily in Britain and, to a lesser extent, in the United States. It is of interest that Keble College Oxford and its chapel, cradle of the Oxford Movement and the home of Holman Hunt's famous painting, *The Light of the World*, was built mainly with the money of the Gibbs family, obtained through commerce in guano. The family built a house called Tyntesfield, near Bristol in the south of England (figure 3.11), to the highest artistic and technical standards of the Victorian era. It and its contents have survived almost unchanged to this day.

Importation of guano into Britain began around 1820, and by 1858 Britain was importing 300,000 tons per year (U.S. imports peaked at about 175,000 tons) and guano was the major source of income of the Peruvian government. Although some of the income was used in development, such as for railway building, most of it went to military and illegal uses. It is recorded that, in 1860, 433 vessels loaded guano at the Chincha Islands, once used by the Incas as their source of guano, and the guano monopoly brought in revenue to the state of nearly 15 million dollars (U.S.). In 1850, guano constituted more than 40% of the United States' consumption of "artificial" fertiliser, at a price of about $73 per ton. In 1869, a French company, Dreyfus, undertook to purchase 2 million tons of guano in exchange for exclusive world rights for its sale. The concomitant financial arrangements restored Peru's credit rating, though it deprived the local businessmen, who had formerly benefited from the trade, of a considerable income.

The demand for guano and the pressure to control its trade were so great that the U.S. Government passed the Guano Islands Act in 1856, which empowered U.S. citizens to take possession of any island, rock, or key with guano deposits not under the control of a foreign government. Such islands included Jarvis Island, Christmas Island, and Midway Island, and many (but not all) of them have stayed under U.S. control until this day.

The Peruvian government effectively mortgaged its future against expected income from sales of guano, but a new competitor, sodium nitrate, was coming on the scene. Britain imported about 150,000 tons of guano per year during the 1860s, but by 1885 the trade and the deposits of guano had virtually disappeared. By the end of that decade, British imports were less than 20,000 tons per year. The trade in nitrate expanded rapidly as that of guano declined. That might not have mattered to Peru if nitrate had simply replaced guano as an export, but Peru soon lost not only her income from guano and nitrate but most of the deposits as well. Peru was essentially bankrupt.

Figure 3.11. Tyntesfield, near Bristol, England, built principally from proceeds of the guano trade. The picture shows the West Front. This Victorian Gothic Revival house was designed by John Norton and built between 1863 and 1866 on the foundations of an earlier mansion. The building and its contents, a showpiece of Victorian craftsmanship and design, are now owned by the National Trust for England. Picture reproduced by courtesy of the National Trust Photo Library/ Andrew Butler.

The Nitrate Wars

Nitrate and guano were recognised to be excellent and cheap fertilisers, and their subsequent exploitation stilled for a time the voices in Britain that had predicted the outstripping of food supplies by population growth. The demand for fertiliser promised a steady income to the countries that, formally, owned the resources. What actually happened was quite different.

In 1830, a shipload of 700 tonnes of nitrate (saltpetre) left what was then a Peruvian port in the province of Tarapaca. The Peruvian government was not terribly interested in nitrate, though the United States and Europe already valued it as a source of fertiliser nitrogen and as a constituent of gunpowder. A rich and secure supply of this material was of considerable interest, but the Peruvians were quite happy with the income from guano.

The common boundaries of Peru, Bolivia, and Chile were not the same as they are today, and they are shown in figure 3.12. Peru had a common border with Bolivia, which claimed a coastal region as far south as the twenty-fifth

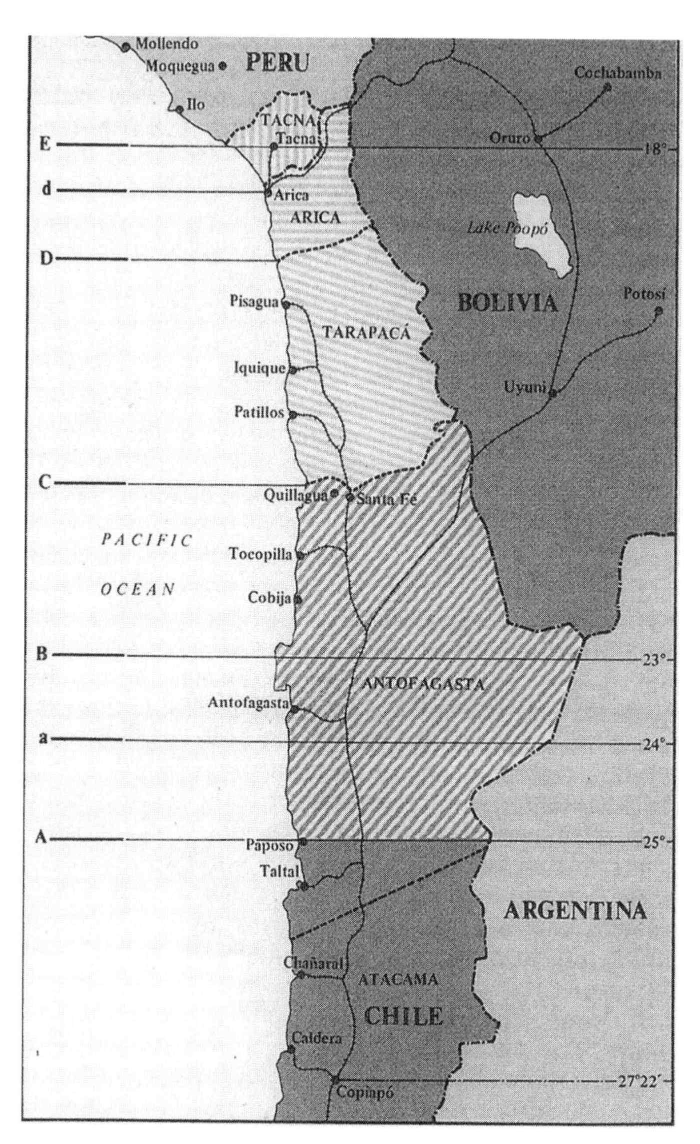

Figure 3.12. The boundaries and claims of Chile, Peru, and Bolivia as they existed at the time of the War of the Pacific. **A**. Original Chile-Bolivia boundary. **B**. Claimed by Chile in 1842. **C**. Original Peru–Bolivia boundary. **D**. Boundary of Chile after the War of the Pacific, 1883, with **D-E** occupied by Chile for 10 years. **d**. Chile–Peru boundary from 1929. Taken from W. J. Dennis, *Tacna and Arica*, Yale University Press, New Haven, Conn., 1929.

parallel, which was its border with Chile. Chile was in dispute with Argentina in Patagonia, a disagreement that was not settled until arbitration by Pope John Paul II in 2000. However, Chile, backed by British commercial interests, was vitally interested in the region to the north of the twenty-fifth parallel that contained deposits of guano, nitrate, and copper, as well as iodine. Figure 3.13 gives some idea of the environment in which were found the deposits of nitrate that were available for extraction. The exploitation of nitrate was controlled principally by Chilean companies backed by British finance. Consequently, there was a continual dispute between Chile and Bolivia concerning the rights to extract and tax the nitrate. Although the Peruvian government had established a monopoly over guano extraction, most of the people controlling and extracting the nitrate, and living in the provinces of Tarapaca (Peru) and around Antofagasta (Bolivia), were not Peruvian or Bolivian. In 1842, the Chilean government claimed much of the Bolivian coastal territory up to the twenty-third parallel, and by a treaty of 1866 it was agreed that the area up to the twenty-fourth parallel should be ceded to Chile. However, the revenues from the whole area between the twenty-third and the twenty-fifth parallels were to be divided between the two countries.

Figure 3.13. The nitrate port of Calita Buena, Chile, as illustrated in Isaiah Bowman, *Desert Trails of Atacama*, American Geographical Society Special Publication 5, 1924. This was a desert port, and the labour conditions must have been extreme. There was a steep descent from the plateau containing the nitrate to the water's edge. Reproduced from the American Geographical Society Library, University of Wisconsin-Milwaukee Libraries.

In 1873, Peru and Bolivia signed a secret alliance of nonaggression and mutual support against third parties (meaning Chile) and undertook not to cede any territory or privileges to anyone else without mutual agreement. By 1875, Peru had realised that its reserves of guano were depleted and had recognised the need to retain control of its nitrate fields. It nationalised them, issuing in return government bonds as surety. Despite pressure upon Chile exercised by the bondholders, some of them even Peruvians, to annex the southern Peruvian province of Tarapaca in order to guarantee some returns, the Chilean government did not then respond. In 1878, Bolivia attempted to impose new taxes on nitrate exported from the province of Antofagasta, and the Chileans then landed troops at Antofagasta City. They also sent a warship to the port of Cobija, "to protect Chilean interests," though this harbour was north of the twenty-third parallel and outside the region where Chile had an accepted interest.

Ultimately, in 1879 a full-scale war, now called The War of the Pacific, broke out. The territories involved were mountainous or desert and not suited to land warfare. The navies were small and not well-equipped or well-manned. There were coups in both Peru and Bolivia, and the Chileans and their backers were victorious. The Chileans pursued the Peruvians far to the north and occupied and sacked the capital, Lima. The library of Lima was despoiled and much of its contents taken back to Santiago, where they remain to this day. The Peruvians finally signed a treaty at Ancon in 1883 by which the province of Tarapaca was ceded to Chile and the provinces even further north, Tacna and Arica, were to be occupied for ten years, at which time a plebiscite was to be held to decide their ultimate fate. The plebiscite was never held, and the matter was settled only in 1929 when the most northern province, Tacna, was returned to Peru, which also was granted some access rights in the port of Arica.

In The War of the Pacific, the Chileans occupied all of the Bolivian Pacific territories. Bolivia effectively withdrew defeated, and thus, as is still true today, Bolivia lost all access to the Pacific, as well as the riches of those territories. A treaty signed in 1884 accorded the Chileans, by an indefinite truce, the right to temporary occupation of the Bolivian littoral. No peace treaty was signed for another twenty years, and then Bolivia surrendered the area of Antofagasta to Chile in perpetuity. It is doubtful whether either Peru or Bolivia has yet become entirely reconciled to these defeats. Bolivia was finally also granted access to Arica, though the town and province remain indisputably under Chilean sovereignty. Arica was declared a free port by the Chilean government in 1953.

The loss of the mineral-rich provinces of the Atacama desert plateau condemned both Peru and Bolivia to poverty, but Chile and the mainly British interests exploited the nitrate deposits to the full. In 1880, there were 2800 workers in the nitrate industry, and the production was 224,000 tonnes, all exported. By 1920, there were 46,200 workers, and exports amounted to 2,794,000 tonnes. This was the peak of the industry, supported by the demand for fertiliser to maintain the burgeoning populations of Europe and North America and for explosives to fight World War I. However, the death throes of the nitrate industry were already evident. In Germany, by 1913 the ultimate solution of the problem of industrial nitrogen fixation had been found, and indeed this helped to support Germany in the war, cut off as it was from the Chilean nitrate. "Synthetic" nitrate production exceeded the Chilean production by the 1930s, and by 1980 Chile produced only 0.14% of the world supply. It is reckoned that Chilean reserves could in 2000 supply world requirements for nitrate for one year only.

Fixed Nitrogen at the End of the Nineteenth Century

The supply of nitrogen from South America allowed Europe and North America to overcome the threat of famine that had so frightened Malthus. However, by the end of the century, the spectre of famine was again worrying some people. The address of Sir William Crookes to the British Association in 1898, described in chapter 1, crystallised ideas that were currently again in circulation concerning the agricultural nitrogen deficit. By then, the problem of fixing nitrogen on an industrial scale was at the forefront of many minds, and it was as much an academic challenge as an economic one. The reality of biological nitrogen fixation by plant systems had by then also finally been unequivocally demonstrated. This will be described in chapter 5.

The fact that in 1898 an eminent scientist not only knew that nitrogen was an essential and limiting nutrient for growing crop plants and knew that nitrogen was one element among many, whereas 300 years earlier scientists were still talking about a mysterious entity called nitre that was different in kind from the four basic elements that were supposed to constitute all matter, is a measure of the scientific and intellectual advances that had been made in those 300 years. This change will form the subject of the next chapter.

CHAPTER 4

The Discovery of Nitrogen and the Disappearance of Alchemical Nitre

The Rise of Agricultural Chemistry in the Eighteenth and Nineteenth Centuries

Nitre, Philosophical and Actual

So far, we have seen how sophisticated systems of agriculture had grown up in many different places and at various times in order to overcome problems associated with the decline of soil fertility arising from continuous exploitation. In Europe and elsewhere, it was clearly understood that manures and various materials such as potassium (or sodium) nitrate could rejuvenate the soil, and empirically probably little more could have been achieved in this direction. Nevertheless, the supply of the products capable of doing this was clearly limited. Only when the scientific basis of the action of fertilisers and manures had been fully understood could further advances be made, and this only happened with the scientific revolution, which began to flower in the sixteenth century and continues in bloom to this day.

The empirical experience of centuries seems to have led to the supposition in Europe that the air was somehow involved in restoring the fertility of soils and in the facilitation of plant growth. However, the reason for this influence could not have been presented in modern terms. A lot of the discussion was centred about the mysterious substance nitre, which was then not simply the salt we recognise today.

There are many instances of statements to the effect that nitre was absorbed from the air and even references in the older literature to aerial nitre. Solid nitre was, of course, very well-known in the form of saltpetre and was widely employed as a constituent of gunpowder. This kind of nitre could also be used as a fertiliser, though there was not enough of it around to "waste" by spreading it on the soil. Then, as is often true today, warfare was regarded as a more important use for such a resource. Nitre could be extracted from manures and from ashes, and, because it was a crystalline solid, it certainly was not the mysterious something that was present in the air. There was no understanding of the modern notions of elements and compounds.

It would take a long time—two centuries—for a truly scientific approach to agricultural chemistry to be developed, but it is still worthwhile to enquire what exactly writers of treatises in the mid-seventeenth century really meant.[1] Clearly, new manures were a matter of considerable interest, and the English patent literature is full of claims for new ones.[2] These patents are often difficult to understand because the modern method of presentation (even today, patents are often framed to conceal as much as they reveal!) had not been developed.

One of the earliest examples of such a patent, John Shawes's English patent number 95, dated 1636, is entitled: "Diverse Wayes and Meanes for the Better Manuring and Improveing of Grounds of all Sort not formerly Found Out nor Practised by Any." No details whatsoever are provided. The patent seems simply to have given Shawe the right to exploit his discovery, whatever it was, and the only indication of the content was in the title. This was apparently normal. Patent number 438, dated 1721, is slightly more informative though also puzzling because the source of its efficacy as a fertiliser is not obvious. The patent claimed a new fertiliser made from chalk and seawater. Even as late as 1773, Baron van Haake in Patent 1049 described a composition that "consists chiefly of common salt, which is melted in an oven (made for that purpose) by a large coal fire until it dissolved as thin and fluid as water; then the same is fixed with saltpetre (which receives from the air a magnetic power, and communicates the same to the fluid) . . . composition becomes of a magnetic quality whereby it attracts its fertility, and is . . . productive."

The Origin of the Term "Nitre"

The concept of nitre goes back a long way, to the time of the alchemists, but understanding the old literature requires more than an ability to realise how the alchemical mind functioned. The *Oxford English Dictionary*[3] states that nitre now means saltpetre but that formerly this material was called natron, which now means a native form of basic sodium carbonate, sodium sesquicarbonate hydrate, such as is found in Egypt and elsewhere. The explanation for this confusion takes us back to the Egyptians and the classical Greeks.

The Egyptian desert contains salt lakes that are the source of large deposits of basic sodium carbonate. The ancient Egyptians used this material for washing and perhaps for making soap, and certainly it was used during mummification processes. The Greeks imported this material from Egypt as a cleaning material, and they called it, in Greek script, νιτρον, or nitron. In 1799, the eminent French chemist C. L. Berthollet, who was a scientific expert for Napoleon's militarily disastrous but scientifically fruitful expedition to Egypt (they unearthed the Rosetta stone, later captured by the British), identified this material as a form of basic sodium carbonate. The word natron itself, the basis of the name of this material in some languages and of the universal element symbol for sodium, Na, passed from Ancient Egyptian or Greek into Arabic, and thence into Spanish and French. According to the *Oxford English Dictionary*, the word natron was first used in an English text in 1684 ("The Natron . . . is an Alkaly Salt perforated like a Sponge, and of a lixivial tast").

In a treatise on the history of alchemy published in 1885,[4] M. Berthelot (a later chemist, not to be confused with Berthollet) adduced a quotation from Plato, who was describing various kinds of "earth" (a name that persists in English in terms such as "rare earth" and "alkaline earth"): "When this earth is freed of most of the water with which it is mixed, it also produces a semi-solid that can be again dissolved in water: thus we can produce on the one hand natron, which can be used to clean oil stains and earth." A footnote by Berthelot states that natron is: "Carbonate de soude: traduit à tort par nitre par les auteurs étrangers à la chimie" (Sodium carbonate: translated in error as nitre by authors who are not familiar with chemistry). Even though Berthelot ascribes this to a careless translator, it may be that the alchemists who first described these salts were not then able to distinguish clearly between the two types of salt. Nevertheless, both words, natron and nitre, seem to have sprung from the same Greek root, νιτρον, nitron.

Just to add to the confusion, a related native hydrated sodium carbonate was also called trona. This seems to have appeared first in English in 1799. The word is also derived from νιτρον but via the Swedish! In 1706, another variant, anatron, this time derived from the Spanish, and also meaning a kind of native sodium carbonate, was described. The use of nitre in the sense of natron in English disappeared about the middle of the seventeenth century when, however, it was still also this mysterious spirit somehow concerned with the air. For example, in 1647 one English writer (presumably of a Calvinist inclination) was describing the time "When God will purge this land with Soap and Nitre, Woe be to the Crowne, woe be to the Mitre." This nitre is presumably basic sodium carbonate. Less than twenty years later, in 1661, another writer was stating: "In the rain, it is not the bare water that fructifies, but a secret spirit or nitre descending with it." This use clearly has little to do with soap. These details are generally taken from the *Oxford English Dictionary*.

Nitre and the Paracelsian Insight

For the origin of the term nitre, or what was apparently the equivalent in many texts, aerial nitre, in this semi-mystical concept, we have to look a little further back in time. It seems to have evolved from Paracelsus (1493–1541) (figure 4.1), who drew a parallel between the heavens (the firmament) and the human body (the firmament within man). Despite a mystical basis for his ideas, Paracelsus was one of the first people to try to use alchemy and chemistry in a scientific fashion.[1, 5] Whereas Paracelsus' ideas involved ideal materials such as sulfur, salt, and mercury that were almost Platonic, he also believed

Figure 4.1. Philippus Aurelius Theophrastus Bombast von Hohenheim in 1538. Not surprisingly, this imposing name was shortened in usual commerce to Paracelsus. He is portrayed at the age of 45, and if this is a genuine portrait then he does not seem to be happy! Several other images look rather similar. Reproduced from an original in the Germanisches Nationalmuseum, Nuremberg, Germany. Another frequent idealised representation of Paracelsus shows him holding a large sword vertically with its tip resting on the ground. The pommel of the sword appears to carry the word "azoth," which bears a resemblance to the current French word for nitrogen, azote. However, it was apparently an alchemical name for mercury and also for Paracelsus' universal chemical remedy for illness, still to be discovered.

that illness indicated a chemical malfunction and that a proper treatment of disease could be achieved through the correct application of chemicals. Modern drug companies would not disagree too strongly!

Paracelsus was an experimentalist, and he established a new tradition of research in chemistry. He attempted to use chemicals as he understood them to treat human disease. He stood at the boundary between the old accepted dogma and the new, much of which he formulated, and some (probably including Paracelsus himself) have likened him to a chemical Luther. In the absence of what we would regard as an adequate model to use in analysing his observations, he was forced to develop his own. As is to be expected, it was derived from the ideas then current and leaned considerably on the older philosophers, though far from uncritically. In particular, Paracelsus stated very clearly that older opinions than his about the universe and medicine were not to be trusted because the ancient philosophers knew no chemistry. He based his unifying ideas upon analogies between, for example, cosmology and medicine, supposed parallels between the heavenly firmament and the internal firmament of the human body. From this standpoint, what one might learn about astronomical theory should also allow you to interpret how the body functions or malfunctions. The interaction between these two firmaments, external and internal, was due to astral emanations, and this idea led eventually to the concept of aerial saltpetre or aerial nitre. This was the basis for the now superseded discipline of iatrochemistry and also of the use of chemicals to treat disease.

The existence of this emanation was supported by the well-known observation that air, an element in the Aristotelian scheme, was necessary to support both fire (another element) and life. It is ironic that a constituent of air really does do this, but it is, of course, dioxygen. However, to the mind of a person such as Paracelsus, the fact that saltpetre could restore soil fertility and also produce the explosion of gunpowder must surely have indicated a common factor here, too. There seemed to be a valid analogy between life and combustion, and they might indeed be essentially the same process. The concept of an aerial nitre responsible for these two similar functions seems not to be at all unreasonable in the light of the knowledge of the time.

Later Paracelsians were to make much use of the idea of aerial nitre, and even Boyle and Newton found the concept fruitful. For example, Sendivogius, a somewhat mysterious alchemist who was, to judge from contemporary reports, also a successful transmuter and who died in 1597, wrote: "Therefore when there is Rain made, it receives from the Air that power of life, and joyns it with the Salt-nitre of the Earth . . . and by how much more abundantly the Beams of the Sun beat upon it, the greater the quantity of saltnitre is made, and by consequence the greater plenty of Corn grows."[6] The concept of nitro-aerial spirit, or aerial nitre, or even aerial saltpetre, seems to have become widely accepted by the end of the sixteenth century and was used throughout the seventeenth century as a useful and generally applicable concept in chemistry, agriculture, and medicine. That it was an ideal Platonic material that could not be isolated was not a problem. It was no different in that way from alchemical ideals such as aerial sulfur and mercury. The concept survived until a truer nature of matter was recognised, but this didn't happen until the eighteenth century.

One may surmise that the use of the term nitre in a common newspaper, as quoted several times in chapter 3, is simple evidence of the widespread adoption of this notion. Robert Fludd (1547–1637), a vigorous propagandist for new philosophy and education,[7] and who was extremely critical of many of the classical ideas then still taught in the universities in many parts of Europe, wrote that "the elementarie aire is full of the influences of life, vegetation, and of the formall seeds of multiplication, which aboundeth with divine beams and heavenly gifts." He seems to have identified this component of "aire" with the breath of the Lord, fire, or saltpetre, in the established manner.

Indeed, Fludd carried out the experiment of burning a candle in an enclosed volume of air until the residual air could no longer support the flame (figure 4.2). The rise of water in the tube containing the candle was clearly observed as the flame burned, but Fludd interpreted his observations in terms of the aerial spirit and did not recognise that a substance had been consumed. His was clearly not yet a rigorously scientific analysis, but it was closer to empirical experience and to experiment than was much of the earlier theorising.

One of the most extensive treatments of aerial nitre and its characteristics was due to John Mayow (figure 4.3), who published a series of five treatises in Oxford in about 1668.[8] One of these was called *De sal-nitro et spiritu nitro-aereo*. Mayow has been rather overlooked in the history of chemical ideas, and probably he was not very innovative, but he was certainly a Paracelsian and thus came before the chemical revolution that revealed the true nature of the elements. Right at the beginning of his treatise, he makes his position clear: "That this air surrounding us . . . is impregnated with a universal salt of a nitro-saline nature . . . will be obvious . . . from what follows." He discusses experiments on nitre that, as far as can be judged, were correctly observed, though not interpreted with present theory. He describes where saltpetre is found or isolated and then states that: "The generally received opinion is that

Figure 4.2. The candle experiment, as portrayed by Fludd. This was originally printed in *Integrum morborum mysterium: sive medicinae catholicae, tomi primi, tractus secundus*, published by Fitzer in Frankfurt in 1631. The figure is reproduced in A. G. Debus, *The Chemical Philosophy, Paracelsian Science and Medicine in the Sixteenth and Seventeenth Centuries*, Science History Publications, New York, 1977, p. 335. It is not evident from this picture exactly how the experiment was set up, but the rise of the water level within the upturned flask is clearly evident. Note that any dioxygen consumed by combustion would be replaced by some of the carbon dioxide generated, so the volume reduction of the flask's gaseous contents would not have been 20%.

Figure 4.3. John Mayow, reproduced courtesy of the Clendening History of Medicine Library, University of Kansas Medical Center.

the earth as its proper matrix draws *sal nitrum* from the air in virtue of its own attractive force." The earth apparently provides something for solid sal nitrum to form, and this something he called a fixed salt: ". . . if earth from which all the nitre has been lixiviated be exposed to air, it will after some lapse of time abound once more in nitre." The reason that ashes and manures are good fertilizers is that they are sources of fixed salt with which the aerial nitre can combine. This clear statement was not really an original opinion but reflected a general belief. Nitre or saltpetre, a solid material that had been known for centuries, was somehow related to an ideal nitre that was in the air and was all-pervasive.

Mayow wrote that the nitro-aerial salt, "whatever it may be, becomes food for fires, and also passes into the blood of animals by means of respiration." In fact, he also performed the famous experiment with a candle in a sealed system that is discussed in more detail below. In addition, he seems to have been aware of an acid derivable from solid saltpetre that was probably nitric or nitrous acid, but his interpretation of the observations was always clouded by his insistence that some form of nitre, with the remarkable properties of supporting life and fire, was drawn from the air. Mayow was also aware of a substance he called Sal-Armoniac, which seems to be present in urine and, he also claimed, in blood. This was very likely a derivative of ammonia.

It is ironic that the agent that Mayow was generally describing was much more like dioxygen than dinitrogen and that the fixing of nitrogen, if that is related to trapping of nitro-aerial particles, was going to be one of the most intensively studied and also one of the most elusive phenomena that has ever been researched. Parenthetically, it seems that the Chinese were aware of the life-supporting agent in air as early as the eighth century A.D., and they described it as the yin of the air.[9]

The Nature of Air and the Discovery of Dinitrogen

It was only when the true nature of air and of the existence and concept of gases was established that a proper scientific understanding of the significance of real material, solid saltpetre, could develop. The concept of a gas is probably due to van Helmont of the famous tree experiment. The history of the unraveling of the constitution of air is somewhat chastening. The simple version is that Joseph Priestley in 1774 or 1775 again burned a candle in a closed bell jar and noted that once some 20% of the air was used up, the candle would no longer burn. By this experiment, Priestley is often said to have discovered oxygen (or, more properly, dioxygen, O_2). The gaseous residue then turned out to be nitrogen (or dinitrogen). In fact, many others were working at that time on related research, and it is not evident that Priestley at first clearly understood the significance of his findings. Although Priestley did indeed carry out the candle experiment, others, including van Helmont and Fludd, as mentioned above, and maybe even Hero of Alexandria in the first century A.D., had also performed it and made similar observations. However, they had interpreted it differently. The reason why Priestley made different inferences was because the intellectual climate of the time was now different, and it was no longer fashionable to employ the semi-mystical concepts of the previous century. A Swedish apothecary, Karl Wilhelm Scheele, also has a claim to be the discoverer of dioxygen, though he no more than Priestley understood the nature of what he had prepared.

The fact that common air was composed of at least two kinds of "aire" was apparently recognised by Joseph Black in about 1756. He obtained a lot of evidence for the existence of a "fixed air" that, with hindsight, was clearly carbon dioxide. In 1722, Stephen Hales, a minister from Teddington, near London, carried out a series of destructive distillations of vegetable matter and collected the resultant gases over water for the first time (figure 4.4).[10] Hales also obtained dioxygen by heating mercury oxide, though he could not have understood his observations in these terms. Priestley, who was also a pastor, extended this gas-collection technique but using mercury rather than water. It thus became possible to trap and investigate gases.[11]

At that time, the favoured explanation of combustion was in terms of a combustible substance containing a species called phlogiston. When some-

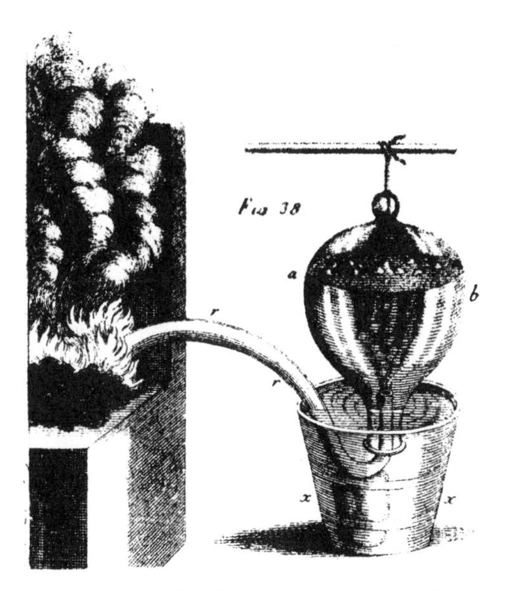

Figure 4.4. Hales apparatus for collecting gases over water. This was an enormous advance because it enabled "philosophers" to investigate and recognise the different kinds of gas. The tube in the centre of this sketch collects hot gases from the destructive distillation of vegetable matter and leads them to the upturned flask, initially completely occupied by water. The accumulating gases would expel the water from the flask. The original was published in Hales's book, *Vegetable Staticks*, of 1727, and this version is taken from A. E. Clark-Kennedy, *Stephen Hales, D.D., F.R.S.*, Cambridge University Press, Cambridge, 1929.

thing burned, it gave up its phlogiston, though several "philosophers" realised that, when something burned, the residue was generally heavier than the starting material. Therefore, phlogiston should have had a negative weight, and this was indeed believed in some quarters. Priestley was convinced of the existence of phlogiston until he died. He involved it in explanations of his observations. It was Lavoisier who seems to have been the first to recognise the true nature of dioxygen.[11]

Perhaps more important was the fact that chemists were beginning to develop a theory of matter that was not simply an extension of medicine (a legacy of Paracelsus) and were beginning to discard the ancient ideas of elements in favour of a more empirical approach that maintained that only a material that cannot be further broken down is really an element. A corpuscular theory of matter, a forerunner of the atomic theory promulgated in its ultimate form by John Dalton in 1805, was also in circulation, initially due to Boyle (whose book *The Sceptical Chymist* was published in 1661)[12] and, through him, perhaps to Newton.[13] Boyle's Proposition I in *The Sceptical Chymist* is as follows: "It seems not absurd to conceive that the first production of mixt bodies, the universal matter whereof they among other parts of the universe consisted, was actually divided into little particles of several sizes and shapes variously moved." His Proposition II seems to anticipate the later definitions of elements and compounds: "Neither is it impossible that of these minute particles divers of the smallest and neighbouring ones were here and there associated into minute masses or clusters . . . not easily dissipable into such particles as composed them." Boyle, amongst others, also recognised that air is a mixture ("a confused aggregate of effluviums"). The ground was slowly being prepared for the development of modern atomic theory. However, Boyle also discussed the "notable practice of the boylers of salt-petre, who unanimously observe . . . that if earth pregnant with nitre be deprived, by affusion of water, of all its true and dissoluble salt, yet the earth will after some yeers yield the salt-petre again; . . . the seminal principle of nitre latent in the earth . . . transforme(s) the neighbouring matter into a nitrous body; though I deny that some volatile nitre may . . . be attracted out of the air."

It is important to realise that none of the great figures of chemical history was operating in an ideological vacuum. Many conflicting sets of ideas were in circulation at the same time, and what we now regard as "correct" or "true" interpretations of nature were often accepted only slowly. The scientists of the seventeenth and eighteenth centuries, like present-day academics, corresponded with each other and exchanged and published information, and every advance, however significant, inevitably owed its origins to the combined efforts of the scientific community.

The discovery of nitrogen in 1772 is generally ascribed to a Scottish student of Joseph Black named Daniel Rutherford. He followed Black in burning organic matter in a fixed volume of common air until no more combustion was supported and then removing the "fixed air" (carbon dioxide) with alkali. He wrote: "By the respiration of animals, healthy air is not merely rendered mephitic, but it also suffers another change, for after the mephitic portion is absorbed by a solution of caustic alkali, the remaining portion is rendered salubrious; and although it occasions no precipitate in lime water, it nevertheless extinguishes flames, and destroys life." (It seems that the word salubrious did not mean exactly what it means today!) The "mephitic gas" that was left was termed by Rutherford phlogisticated air. Earlier workers who had also observed a reduction in gas volume in such experiments had not investigated fully the volumes and nature of the products. That reduction in volume could not have been the 20% directly associated with dioxygen because combustion products took its place to some degree, and the temperature within the closed vessel must also have changed. Today, we recognise that Rutherford used up all the dioxygen in the air by combustion, absorbed the resulting carbon dioxide in alkali, forming a carbonate, and was left with a gas that was mainly dinitrogen. Again, it was probably Lavoisier who first interpreted such observations in modern terms, though whether it is correct to describe him as the discoverer of dinitrogen is, at the least, questionable.

Henry Cavendish performed essentially the same experiment, though more carefully than Rutherford, and he measured the volumes of gases involved. He distinguished this mephitic air (dinitrogen) from fixed air (carbon dioxide). He informed Joseph Priestley of his observations. Scheele had also separated air into its two major constituents, as did Priestley in 1774. It was becoming evident that common air contained a gas that supported life and combustion and at least one other gas that did not. Nevertheless, the component of air closest in properties to the aerial nitre of old was not dinitrogen but dioxygen.

Priestley and Scheele developed several methods of producing dioxygen, and Lavoisier, who first recognised the true nature of combustion, was aware of at least some of these. Priestley informed Lavoisier that heating calx of mercury (mercury oxide) generated "dephlogisticated air" (dioxygen), and this was confirmed by Lavoisier. In 1778 Lavoisier announced that this dephlogisticated air could oxidise a metal and that this air also combined with charcoal to generate the fixed air (carbon dioxide) of Black. Lavoisier was notable for making quantitative measurements of the weights of his solid materials, though it has been claimed that the increase in weight of a metal upon calcination (oxidation) was noted first by Jean Rey. However, Rey's pamphlet of 1630 makes it clear that he was repeating and rationalising (wrongly, of course) the observations of even earlier experimenters. By the efforts of numerous workers the nature of dioxygen was being uncovered, and that of dinitrogen almost as an aside.

Lavoisier had observed that both sulfur and phosphorus could burn in air to produce species that dissolved in water to produce acids. This, though now known to be misguided, is the basis for the name "oxygen," or acid producer, which has the precise equivalent, *Sauerstoff*, in German. Lavoisier called the residual, mephitic air *la moufette atmosphérique*, and the German equivalent, due to Scheele, was *die verdorbene Luft*. The modern versions have comparable meanings, French *azote* (without life) and German *Stickstoff* (suffocating material). The present-day name in English, nitrogen, or source of nitre, was coined a little later.

The identity of nitre, sodium nitrate, and characteristics peculiar to it were not really recognised by the alchemists of Western Europe prior to the thirteenth century. Apparently, nitric acid was described by the Arab chemist Geber (Jābir ibn Hayyān) of the eighth century A.D., to whom a preparative method was ascribed in a twelfth-century Latin manuscript. The recipe is as follows: "Take a pound of Cyprus vitriol, a pound and a half of saltpetre, and a quarter of a pound of alum. Submit the whole to distillation in order to withdraw a liquor which has a high solvent action." This product is indeed nitric acid, then known variously as *aqua fortis*, as well as *aqua dissolutive*, *aqua acuta*, *aqua calcinativa*, or *aqua valens*. Its formation from a sulfate, a nitrate, and a hydrated alum is easily understood in modern terms. Geber even notes that if sal ammoniac is added to this mixture, then *aqua regia*, a solvent even for gold, is obtained. In 1658, Johann Rudolf Glauber described a much more detailed method to prepare nitric acid from sulfuric acid and nitre, a method that might still be used today in certain circumstances. Consequently, a different nomenclature grew up, including *spiritus nitri fumans Glauberi*, *acidum nitri*, and *acide nitreux* (nitrous acid). It took a considerable time before all these names were understood to designate essentially the same material, and this must have caused some confusion. The name nitric acid only became current after 1787, when the name *acide nitrique* (nitric acid) was generated in one of the first modern attempts to produce a systematic inorganic chemical nomenclature.[14]

Lavoisier had established as early as 1776 that nitric acid contains oxygen. This was consistent with his oxygen theory of acidity. Cavendish described experiments that involved subjecting mixtures of dinitrogen and dioxygen (not that he called them such) to an electric spark. The water that was used to close the vessel containing these gases was shown to contain nitric acid. "We may safely conclude that in the present experiments the phlogisticated air [i.e., dinitrogen] was enabled, by means of the electric spark, to unite to, or form a chemical combination with, the dephlogisticated air [i.e., dioxygen], and was thereby reduced to nitrous acid, which united to the soap lees [effectively, sodium hydroxide or carbonate], and formed a solution of nitre." This extraordinary experiment, though then interpreted in terms of the now defunct phlogiston theory, is the first account of the fixation of nitrogen. It established the connection between saltpetre and atmospheric nitrogen and showed that the salt contains both oxygen and nitrogen. One of the industrial methods, now defunct, to fix nitrogen employed essentially the same method. In 1790, Jean-Antoine Chaptal suggested the name nitrogen, which indicates its relation to nitre. The precise composition of nitric acid was established only in 1816 by Joseph-Louis Gay-Lussac and C. L. Berthollet.[15]

The final key chemical in this story is ammonia. The name itself has a long history. In about 440 B.C., Herodotus wrote a history that describes a sandy ridge in Libya that extends from Thebes in Egypt to the Pillars of Hercules (the Strait of Gibraltar).[16] The first stop along this ridge is where the Ammonians live: "At intervals of about ten days journey along this ridge are masses of great lumps of salt in hillocks; on top of every hillock a fountain of cold sweet water shoots up from the middle of the salt hillocks." This spring ran cold at midday and conveniently hot at night. Whatever the true identity of this fabulous material, which some people used as building blocks for houses because in that region it never rains, it seems to be the ultimate source of the name ammonia. Herodotus' view of the local geography of the area is shown in figure 4.5.

Much later, Pliny the Elder (A.D. 23–79)[17] refers to Hammoniac salt "so-called because it is found under the sand . . .", though he also states that it is also found "as far as the oracle of Hamman" in Africa. Perhaps the derivation is also related to the name of the Egyptian god Amun. Pliny notes elsewhere that the calcinations of nitrum produce a "vehement odour," which seems to confirm that different salts were often confused, and the identity of this material, as distinct from, say, rock salt, sodium chloride, seems not to have been

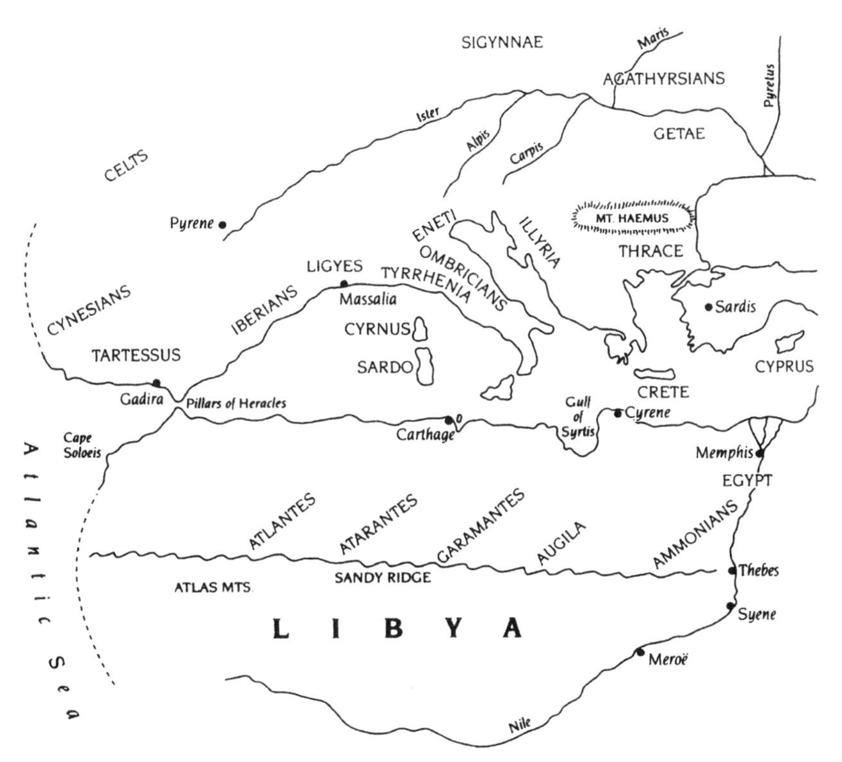

Figure 4.5. The world according to Herodotus. This map is adapted from Herodotus, *The History*, volume 2, book IV, translated by A. D. Godley, The Loeb Classical Library, William Heinemann, London, 1921, reprinted 1957. The map shows where the Ammonians were supposed to live. Notice also that the Celts, now restricted to the western fringes of Europe, once apparently occupied much of France.

clearly evident at that time. Nevertheless, the material sal ammoniac (now ammonium chloride, NH_4Cl) was recognised by the Arab authorities on alchemy, and this passed into mediaeval European usage, though sometimes the term was transmuted to "sal armenium." Geber described the preparation of sal ammoniac from an awful mixture of human urine, human sweat, common salt, and wood soot, heated together until the ammonium chloride sublimed out. Camel urine was also used, probably because it was more easily available in quantity, but nothing is mentioned about camel sweat!

The method of preparation was refined through the centuries, but the Middle East remained a major source for the European market. Boyle apparently stated that it would be cheaper to make it at home rather than import it, and he was prepared to provide the recipe, which probably did not involve camel products. The uses of ammonium chloride were various, including the protection of copper surfaces from oxidation when they were being tinned, and in dyeing.

Another recipe for sal ammoniac, ascribed to Raymond Lully in the fourteenth century,[15] recommended taking a "goodly quantity of the early morning urine of boys from eight to twelve years of age," though this would seem to have been a source of ammonia solution, ammonium hydroxide (NH_4OH), rather than the chloride. There might have been a problem in finding enough boys of the required age, but by the 1750s there were several commercial producers of ammonium chloride in Europe, and the general relationship of ammonia, its aqueous solution, ammonium hydroxide, and its salt with hydrochloric acid, HCl, was being established. Priestley described ammonia itself, a gas under normal conditions, in 1774.

C. L. Berthollet established the composition of ammonia for the first time. He decomposed it by means of an electric spark into three parts of dihydrogen (H_2) and one part of dinitrogen, corresponding to the formula NH_3. It was then clear that ammonia contained nitrogen, and the theoretical relationship to nitre and nitrogen gas must have been evident. Ammonia can burn in air and can even support a flame to some degree, as reported by Berzelius, but I. Milner showed as early as 1789 that ammonia could be oxidised at red heat in the presence of manganese dioxide to yield materials that dissolved in water to give a mixture of nitrous and nitric acids (HNO_2 and HNO_3, respectively). As he reported (in the first person, though not yet in the usual passive voice now common in scientific reports): "I rammed a gun-barrel full of powdered manganese; and to one end of the tube I applied a small retort, containing the caustic volatile caustic alkali [ammonia solution]. As soon as the manganese was heated red-hot, a lighted candle was placed under the retort, and the vapour of boiling volatile alkali forced through the gun-barrel. Symptoms of nitrous fumes and nitrous air soon discovered themselves . . . The receivers are presently filled with white clouds of nitrous ammoniac."[15]

In summary, the following relationships and conversions (figure 4.6) were recognised by about 1800. The compounds were all well-characterised, and the value of some of them, such as nitrate, or of their precursors, such as dung

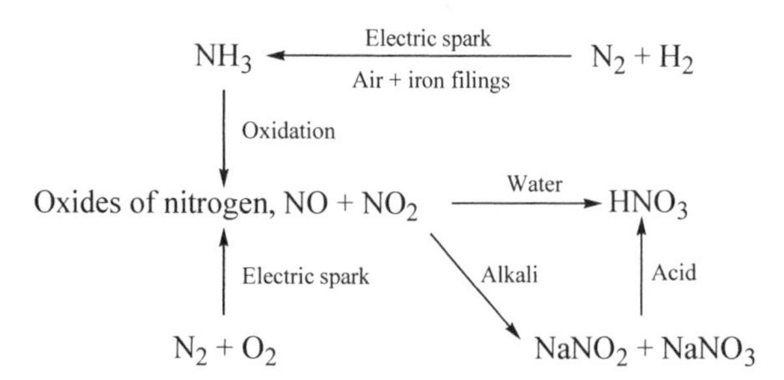

Figure 4.6. The interconversions of dinitrogen and related compounds as established by about 1800. The formulae are modern and would not then have been recognised for what they are. It is notable that the industrial exploitation of this chemistry was not achieved until a century later. The technical means of doing so on an industrial scale had yet to be developed.

and urine, for fertilising crops was a matter of common knowledge. The involvement of nitrogen as a constituent of plant and animal matter was also becoming clear, but the detailed interconversions, in vivo in some cases, were not understood then, and some have yet to be completely understood even today.

However, the same period, the second half of the nineteenth century, saw the emergence of a new branch of chemistry, as influential in its way as the chemistry associated with medicine (iatrochemistry) that was stimulated by Paracelsus, and led to the establishment of chemistry as a modern science. In many ways, the new agricultural chemistry acted as a principal focus for the emergence of biological chemistry.

The Birth of the Scientific Chemistry of Plants

The application of salts of various kinds to soils in order to improve them and to increase soil fertility had been practised long before 1800. However, it was only by about 1800 that it became possible to understand what was really going on, and this then led inexorably to the current widespread application of synthetic chemical fertilisers. However, it should be remembered that once a fertiliser is applied to the soil, the plant absorbs nitrate (or whatever) and it has no mechanism for detecting its origin, whether industrial or "organic." Even if the application of industrial fertiliser were to be completely suppressed, a growing crop would still require a similar quantity of "organic" fertiliser. The long-term environmental effects might not be very different in either case, depending on the relative rates of nitrogen release from the fertilisers.

Amongst the new, patented fertilisers that appeared in the seventeenth century, we have already mentioned a concoction of salt and lime. This patent claim was not as peculiar as it may now seem. For centuries, the value of marl (a mixture of clay and lime) as a soil improver had been recognised. An English chemist named Hugh Plat, resident in the no doubt then leafy village of Bethnel Green (now presumably a part of London's rather less countrified East End and known today as Bethnal Green), wrote a book first published in 1594 and called *The Jewell House of Art and Nature*. Part of it deals with "diuerse new sorts of Soyle not yet brought into publique use" and refers to "all sorts and kinds of Marl, or soyl whatsoever, either known or used already for the manuring or bettering of all hungry and barren grounds . . . draw their fructifying vertue from that vegetative salt." The vegetative salt seems somehow to be a cousin of alchemical nitre. As noted by Debus,[1] this is a rather Paracelsian viewpoint, but there is no doubt that the lime in the marl, at the very least, would have reduced soil acidity. The approach was still empirical and not very scientific.

One of Plat's conclusions appears to be that salt from the sea possessed enough of the virtue of the vegetative salt to be useful itself as a fertiliser. There were even experiments that proved that sea salt could be a better fertiliser than manure. While not doubting the observations, the explanation could be one of many, not the least concerning trace elements in the salt and the age and history of the manure. It had also been observed that fields flooded with seawater did not lose their fertility permanently—rather the converse—and English West Country farmers were known to fertilise their fields with salt soil from the sea.

Samuel Hartlib[18] also recommended the application of marl. He was convinced of the value of rainwater because "we can be certain that it contains a vital salt." However, he was also clear that what all the available fertilising materials actually did was not at all understood. In 1675, Daniel Coxe[19] (the title page of his paper describes him as "the learned and intelligent Daniel Coxe") published a paper on the use of sea sand as a fertiliser and recommended that only damp sand be used, dry sand being of no value. This Coxe was the founder of the American Coxe dynasty and was actually made the governor of West Jersey, though apparently he never visited his estates in America. He seemed to recognise the value of the sand rather than the salt, but the greater significance was that he no longer ascribed all the virtues to some Platonic salt, the existence of very different kinds of salts by then becoming recognised. However, it is clear that a more informed approach to plant nutrition could not develop until the chemistry associated with plants had become more fully developed.

In about 1800, a celebrated chemist, Thomas Thomson, M.D., published a massive four-volume work called *A System of Chemistry*.[20] Thomson was a lecturer in chemistry at the University of Edinburgh, and it is perhaps not an accident that his principal academic qualification seems to have been medical. In his book, the emergence of a scientific approach to the analysis of matter is clearly evident. The book went through at least six editions in twenty years, and it was extensively revised so that the sixth edition presented quite a different picture from the first edition of many aspects of the subject. Chemistry then covered a vast range of topics, and much of it would today be considered biochemistry and biology. Thomson states that: "It is the business of the chemist to analyse vegetables, to discover the substances of which they are composed, to examine the nature of these substances." In volume 4, Thomson discusses the van Helmont tree-growing experiment described in this book in chapter 3. He states that plants consist of vegetable principles, of which he lists twenty-three. These include sugar, gum, starch, and so on. The chemical nature of these principles was then not really known, but the major constituents are stated to be carbon, hydrogen, oxygen, and azote. The last is not present in all the principles, and when it is present it is generally in small proportion. However, Thomson also mentions that plants contain alkalies, earths, and metals. The quantities of these constituents were usually so small that they could scarcely be classed as vegetable principles, but the important point is that they were recognised as plant materials. Thomson believed that they were taken up into plants from the soil and through the roots, and in solution, which was an opinion that had been current for at least two centuries.

Thomson's criticism of the original interpretation of the van Helmont experiment is very much to the point. First, the rainwater employed by van Helmont must have contained dissolved earths. He quotes earlier experiments that showed that one pound of rainwater contains one gramme of earth (or solid residue). This must invalidate any firm conclusion that the plant somehow converted water into earth. In addition, the unglazed vessel in which van Helmont apparently grew his tree would have absorbed earth-containing water from the soil upon which it stood. Thomson quotes other reports that showed that plants grown in pure water from bulbs or from types of bean do not increase in their total content of carbon. This is rather puzzling in the light of our present understanding since photosynthesis should definitely lead to an increase in the content of carbon from carbon dioxide in the air. In fact, the sixth edition of Thomson's book notes reports that contradict these observations. Nevertheless, on grounds such as these, no firm deductions can be made from van Helmont's data.

A brief discussion of fertilisers in Thomson's book in a section devoted to the "Food of Plants" points out that carbon must be in an appropriate form to be taken up—powdered pit coal, for example, being of little value. Experiments with fermented and unfermented dungs showed clearly that carbon must be in an appropriate state of combination ("whatever that may be") and in solution (but not as dissolved carbon dioxide) to be of value to the plant. Plant and animal matter may be useful as fertilisers, but only once they have decayed. The possibility that azote might be a significant factor in all this is not broached.

However, this was not all. Thomson describes experiments that showed that plants produce water from their leaves, a process now termed transpiration. In fact, Joseph Priestley was the first to show (in 1771) that plant leaves absorb carbonic acid gas (fixed air, or carbon dioxide). Though some of his conclusions were later challenged, other workers confirmed this absorption. At about the same time, it was discovered that light is necessary for the absorption to occur and that dioxygen is evolved as the carbon dioxide is taken up. Light also helps leaves to turn green. All these reactions are now realised to be involved in the processes of photosynthesis, and Thomson remarks that this must be a way for plants to accumulate large amounts of carbon.

Thomson also states that during the night plants absorb dioxygen and produce carbon dioxide. This again is now well-understood to be a result of plant respiration. During the night, in the dark, leaves of green plants absorb dioxygen, and during the day, and in sunlight, they absorb fixed air (carbon dioxide). In fact, says Thomson, the processes of gas uptake and evolution are not in balance, but the apparent contradiction between these observations and the bulb and bean results originally quoted above seems not to have occurred to him.

What also seems to be evident from this discussion is that azote apparently has no effect upon plants. It will not support plant growth (as with animals, it seems to suffocate them), and plants do not produce it. Chemists already knew by about 1800 that dinitrogen was a very inert and unreactive substance, and the possibility of its significance for biology was not even considered. The suggestion that plants might be required to do, and somehow were able to do, what Dr. Priestley had been able to achieve only with an electric spark was simply not conceivable. Even the scientific lion of the age, Humphrey Davy, did not consider this likely. Nevertheless, the mystical connotations of nitre and saltpetre were sinking into the unscientific past.

This is not to say that Thomson presents a simplified picture of what one would accept today. In volume 2 of *A System of Chemistry*, he states that nothing "has excited the attention of chemical philosophers more than the continual reproduction of nitre in certain places after it has been extracted from them. Prodigious quantities of this salt are necessary for the purposes of war." He suggests a provocative solution as to how this reproduction occurs: "Several French philosophers . . . discovered that nothing else is necessary for the production of nitre but a basis of lime, heat, and an open, but not too free communication with dry atmospheric air . . . the acid is first formed, and afterwards the alkali makes its appearance." He rightly notes that: "The appearance of potass is . . . extraordinary. If anything can give countenance to the hypothesis that potass is composed of lime and azote, it is this singular fact."

We now know that what Thomson called potass was probably impure potassium carbonate. If his hypothesis were correct, then these observations could be an example of the laboratory or industrial fixation of nitrogen, but it seems doubtful. However, it might also explain the efficacy of the new fertiliser patented in 1721, and mentioned above, that was made simply by melting together common salt and lime. Be all that as it may, a certain Dr. Austin is reported by Thomson to have mixed dihydrogen and "azotic gas" and to have tried to make them combine to give ammonia. He applied electricity, heat, and cold. He even succeeded in 1788 in the following way. "He introduced into a glass tube filled with mercury a little azotic gas, and then put into the gas some iron filings moistened with water . . . the hydrogen of the water meeting with the azote at the moment of its admission (and formed by the action of the iron on the water), combines with it, and forms ammonia . . . This experiment shows that the gaseous state of azote does not prevent its combination with hydrogen." Hence, by about 1800, at least two general methods of fixing nitrogen—oxidative and reductive—were recorded, and both accounts are probably reliable. It was to take another hundred years before these methods could be turned into industrial processes.

The Birth of Agricultural Chemistry

Humphrey Davy

Sir Humphrey Davy was interested in an enormous variety of subjects, though in the popular mind he is remembered chiefly for the miner's safety lamp and for the social use of laughing gas (nitrous oxide, or nitrogen(I) oxide,

N_2O). He was also very interested in agriculture, and he delivered a series of annual lectures to the British Board of Agriculture. These were gathered together and published as a book in 1813. His book presents a discussion of various "scientific" techniques for improving agricultural productivity, including the use of fertilisers.[21]

By 1813, things had clearly moved on from the time when Thomson was first writing in about 1800. Davy discusses albumen, which, to a modern reader, seems to mean protein. In the third lecture, he states that 100 parts of albumen contain 53 parts of carbon, 24 parts of oxygen, 7.5 parts of hydrogen, and 16 parts of azote. Indeed, it is the only vegetable principle that was supposed to contain nitrogen, and Davy states that plant albumen is much less abundant than animal albumen. In lecture 8, Davy makes the statement, based upon chemical analysis, that peas and beans contain a small quantity of matter analogous to albumen. He then states unexpectedly ". . . but it seems that the azote which forms a constituent part of this matter is derived from the atmosphere". Quite how firmly Davy held to this opinion, or even what it implies, is not clear, and no evidence to support this conjecture is provided. His previous discussions on uptake of gases by plants all refer to carbon dioxide, so he could have been implying, as did Justus von Liebig later, that ammonia is similarly absorbed rather than that nitrogen is fixed.

Lecture 6 discusses different manures, including the use of whole animals as fertilisers and even the use of pilchards in Cornwall. Davy lists the use of night soil by the Chinese, urine of various kinds, and the dung of seabirds, which contains ammonia and urea, though "The dung of sea birds has, I believe, never been used as a manure in this country." This statement must have been made just before the start of the importation of guano for use as fertiliser. He warns against excessive fermentation of manure because this causes the loss of carbon as well as some ammonia.

An explanation of the value of manure seems to require a more detailed understanding of plant constituents. In lecture 7, we learn that green plants contain earthy and saline matters not derived from the seeds from which they spring and that these are probably not derived from air and water. It seems ". . . fair to conclude that, the different earths and saline substances found in the organs of plants, are supplied by the soils in which they grow; and in no cases composed by new arrangements of the elements of air and water." This leads to a discussion of manures that ". . . are not necessarily the result of decayed organized bodies which are not composed of different proportions of carbon, hydrogene, oxygene, and azote." The listed inorganic fertiliser materials include quick and slaked lime, caustic magnesia, gypsum, phosphate of lime, and solutions of sulfate, nitrate, muriate (chloride), and carbonate of ammonia. Ammoniacal compounds are, in general, good fertilisers.

Davy was a Cornishman, and he refers more than once to his native county: "Refuse salt in Cornwall, which however, likewise contains some of the oil and exuriae of fish, has long been known as an admirable manure." He is more sceptical about "nitre, or the nitrous acid combined with potassa.

Sir Kenelm Digby states, that he made barley grow very luxuriantly by watering it with a very weak solution of nitre: but he is too speculative a writer to awaken confidence in his results . . . it may furnish azote to form albumen or gluten, . . . but the nitrous salts are too valuable . . . to be used as manure."

In lecture 8, Davy discusses fallowing, and his final opinion about nitrogen in plants and dinitrogen in the air seems to be that there is no connection. Since these lectures were delivered at yearly intervals, it is indeed possible that his opinion changed as more data appeared. Certainly, some of his own experimental results on various kinds of fertiliser were inconsistent: "It has been supposed by some writers, that certain principles necessary to fertility are derived from the atmosphere, which are exhausted by a succession of crops, and that these are again supplied during the repose of the land . . . The vague ancient opinion of the use of nitre, and of nitrous salts in vegetation seems to have been one of the principal speculative reasons . . . Nitrous salts are produced during the exposure of soils containing vegetable and animal remains . . . but it is by probably the combination of azote from these remains with oxygene in the atmosphere that the acid is formed; and at the expense of an element, which would otherwise have formed ammonia." This much seems unequivocal: whatever the ancients say, nitrogen does not get into plants from atmospheric nitrogen. And such was Davy's authority that his conclusions were widely believed. Nevertheless, he was one of the first to attempt a general scientific investigation of plants and of fertilisers of many kinds of plant, animal, and mineral origin.

An Early Feminist and Populariser of Science

Davy's opinions were widely accepted and were retailed in other publications. One of the most interesting was published about the same time as Davy's lectures. It is a book called *Conversations in Chemistry* and was written by the wife of a London doctor, Mrs. Jane Marcet.[22] Mrs. Marcet was a true pioneer and a feminist. The book, like the several companion volumes on a variety of subjects, is cast in the form of conversations between an instructor, Mrs. B., who is never further identified, and her two students, Emily and Caroline. It is aimed at an audience chiefly "of the female sex," for Mrs. Marcet was convinced that women could master science equally as well as men. She was a great admirer of Davy, whom she quotes on numerous occasions, but she was also a shrewd judge of the book market. Her book went through many editions over more than twenty years, and it was revised and expanded as new discoveries and ideas became current. The thirteenth edition was published in 1837, and the original publication sits nicely between the ideas of Davy, active at the beginning of the nineteenth century, and the enormously influential Justus von Liebig, who was active from about 1840 and is generally regarded (though perhaps not entirely accurately) as the father of agricultural chemistry.

In one of her lessons with Emily and Caroline, Mrs. B. leads a conversation on nitric and nitrous acids, and remarks: "the celebrated Mr. Cavendish ascertained (only within these last twenty-five years) that nitric acid consists of about 10 parts of nitrogen and 25 parts of oxygen. These principles, in their gaseous state combine at a high temperature; and this may be effected by repeatedly passing the electrical spark." This alarms Caroline: "But in a thunder-storm, when lightning repeatedly passes through them, may it not produce nitric acid?" Mrs. B. reassures her: "There is no danger of it, my dear; the lightning can affect but a very small portion of the atmosphere." This provokes Emily to ask: "But how could nitric acid be known, and used, before the method of combining its constituents was discovered?" Mrs. B. knows all the answers: "Previous to that period the nitric acid was obtained, and it is, indeed, still extracted, for the common purposes of art, from the compound salt which it forms from potash, commonly called *nitre*." This is too much for Caroline, who is ever the thoughtful student: "Why is it so called? Pray, Mrs. B., let these old unmeaning names be entirely given up, by us, at least; and let us call this salt *nitrate of potash*."

This charming exchange is typical of the whole two volumes. However, there is a more serious side to all this. Mrs. B. includes a long lesson on vegetables. It starts with a list of vegetable principles very similar to the list presented by Thomson thirty years before. Emily is quite aware of the superficiality of this kind of analysis: "Yes, Caroline, you have told us what life *does*, but you have not told us what life *is*." Mrs. B. ducks this percipient question but tells her students that ". . . all vegetables are composed of hydrogen, carbon, and oxygen (with a few other occasional ingredients)."

As was then the custom, Mrs. B. instructs her pupils in the rudiments of agriculture: "The soil, which at first view appears to be the aliment of vegetables, is found, on closer investigation, to be little more than the channel through which they receive their nourishment." This provokes Caroline: "And pray what is the use of manure?" Mrs. B.'s answer is that manure contains the partially decomposed principles that have already been listed. The rich mould produced in woodland seems a particularly good nutrient: "This accounts for the plentifulness of the crops in America [meaning, of course, North America and probably New England], where the country was, but a few years since, covered with wood." This may cause some amusement in, say, New Hampshire, which is now extensively covered with new, secondary woodland but was originally extensively cleared of forest for agricultural purposes.

Caroline then wonders why animal manure is actually so much better than plant manure because the latter should contain more vegetable principles. The reason given by Mrs. B. is that the animal wastes contain a much greater proportion of nitrogen. This renders their composition "more complicated, and, consequently, more favourable to decomposition." Mrs. B. even manages to explain why agriculture can increase overall plant productivity, something that is unlikely if "principles" are the limiting factor and are merely circulating through the biosphere and the atmosphere. She lauds the balance of plant and animal kingdoms and neatly tries to avoid further questions by statements such as: "And how very admirable the design of providence who makes every different part of creation thus contribute to the support and renovation of each other." It was to be another fifty years before this comfortable and satisfied summary of nature was finally shown to be inaccurate.

Justus von Liebig

Davy's contributions to agricultural science were seminal, and his book of lectures was widely studied. What is clear at a distance of some 200 years is that Davy fostered a much more experimental approach to problems of agriculture and was, in many ways, the progenitor of modern agricultural science. However, he was still theorising in terms of principles and substances, and though he realised that silica and earths were minor constituents of plants, he still did not think in terms of plants able to synthesise their own constituents. He seems to have accepted that to enter a plant a substance had to be soluble in water, but deeper than that he did not satisfactorily delve. Davy was aware that nitrogen was important, and he may even have suggested on occasion that nitrogen might be fixed biologically, though this is not unequivocal. The main nitrogenous value of manures was said to be the ammonia and the nitrogen-containing principles they supplied. It was only in 1840, with the publication of Justus von Liebig's *Chemistry in Its Application to Agriculture and Physiology*,[23] that agricultural chemistry began to assume its current modern pattern.

Von Liebig's book starts with a discussion of the constituent elements of plants, but he does not mention any of the "principles" described by earlier workers. The constituents are elements, as understood in modern terms. Concerning nitrogen, Liebig says that: "Its principal characteristic is an indifference to all other substances, and an apparent reluctance to enter into combination with them . . . Yet nitrogen is an invariable constituent of plants." Later, von Liebig states that nitrogen and oxygen are the principal constituents of the atmosphere but that carbonic acid (carbon dioxide) and ammonia are also important in the context of plant growth and decay. A major constituent of soils is humus, otherwise decayed vegetable matter, and this is the principal nutrient of plants. For a plant or animal to be healthy, it requires many things, some in small quantities and some on a large scale. For example: "It is quite impossible to mature a plant . . . the solid framework of which contains silicate of potash, without silicic acid and potash."

Von Liebig then describes in detail the constituents of various types of manure. His prejudices are already clear: "The peculiar action, then, of the solid excrements is limited to their inorganic constituents." In human faeces, these include phosphates of lime and magnesia, and other dungs supply other inorganic matter. Bone meal contains more than 50% phosphates of lime and magnesia. Von Liebig adds: "A time will come when fields will be manured with a solution of glass (silicate of potash), with the ashes of burnt straw, and with the salts of phosphoric acid, prepared in chemical manufacturies."

However, von Liebig also recognised that "every part of the organism of a plant contains azotised matter in very varying proportions," and the ultimate source of this material would seem to be humus. Nevertheless, he continues: "Nature, by means of the atmosphere, furnishes nitrogen to a plant in a quantity sufficient for its normal growth." This statement could indeed be taken to mean that plants could fix atmospheric nitrogen, but that was not the intention. The theory of von Liebig was as follows. Urine and dung are prime sources of nitrogen in the form of ammonia, produced during their putrefaction. This ammonia is volatile if the environment is alkaline, and it is lost to the atmosphere. However, the ammonia can be neutralised by a variety of materials, including gypsum, sulfuric or muriatic (hydrochloric) acids, and superphosphate of lime, producing involatile ammonium salts. This is an explanation of the value of gypsum in fertilising soil, though other explanations are current today.

The general use of night soil is also mentioned by von Liebig, with due acknowledgement to the prior example of Chinese practice: "The Chinese are the most admirable gardeners and trainers of plants, for each of which they understand how to prepare and apply the best-adapted manure. The agriculture of their country is the most perfect in the world." Doctor King (see chapter 2) would have approved of that statement and may even have read it. The preparation of manures from night soil as carried out in London is then described. This is the kind of industry with which Sir William Crookes was later concerned. Finally, von Liebig quotes both Garcilaso de la Vega and von Humboldt as to the origins, uses, and efficacy of guano. As for artificial manures, von Liebig, perhaps echoing Davy, doubts the value of nitrate of soda: "There is nothing opposed to the supposition that nitric acid may be decomposed by plants, and its nitrogen assimilated." After all, he probably reasoned, plants decompose carbonic acid and assimilate the carbon. However, von Liebig knew that growth trials with nitrate of soda were not consistently good, and in some cases the salt seemed to have no effect at all, even when the related potassium nitrate proved to be helpful. This matter was left undecided.

Von Liebig describes how putrefaction of nitrogen-containing plant and animal materials generally releases ammonia to the atmosphere. Ever the experimentalist, he describes how the presence of ammonia was detected in rainwater falling near his laboratory at Giessen. This had not been detected before because no one had looked for it. Von Liebig stated: "All the rain-water employed in this inquiry was collected 600 paces south-west of Giessen whilst the wind was blowing in the direction of the town [to avoid contamination from the town itself]. When several hundred pounds of it were distilled in a copper still, and the first two or three pounds evaporated with the addition of a little muriatic acid, a very distinct crystallisation of sal-ammoniac was obtained." His calculations based upon these observations were taken to show that there is enough ammonia in rain-water or in snow to provide all the nitrogen required by growing plants and thence, by digestion, by animals. Ammonia might well have been the aerial nitre of Houghton's speculations (chapter 3), but von Liebig was unfortunately mistaken in some of his conclusions.

The recognition that rain might contain fixed nitrogen is really the first formulation of what we today call a Nitrogen Cycle (see chapter 1, figure 1.8), the continual circulation of nitrogen (according to von Liebig, as ammonia) from the atmosphere into plants and animals, where it combines to form gluten, albumen, and other substances. Upon death and putrefaction, ammonia is regenerated and released into the atmosphere, only to dissolve in the rain and be assimilated once more. This is an elegant theory and was clearly an improvement on any advanced earlier. Von Liebig was convinced of the importance of nitrogen for plant growth, though having visited some of the leading experimental agriculturalists of the day in England, he later became of the opinion that phosphorus was even more significant. Such was von Liebig's authority, and his aggressiveness when it was challenged, that his ideas, good as well as bad, were widely adopted. Just like Davy, he did not believe that dinitrogen could be converted to ammonia by plants or animals. It seemed to be just too unreactive. He undoubtedly showed the correct way to study the effects of chemicals on plants, and he laid the basis for modern intensive agriculture. But in the matter of ammonia and nitrogen, he was wrong.

The Reality of Biological Nitrogen Fixation

The possibility that plants might somehow be able to fix nitrogen had been raised ever since the requirement of growing plants for nitrogen had been generally realised at the beginning of the nineteenth century. This was at least a question that could be, in theory, solved by experiment, and it was couched in real terms rather than beneath the mantle of some semi-mystical nitre. Nevertheless, the major chemistry authorities of the day, principal amongst them von Liebig, were of the opinion that dinitrogen is so unreactive that it seemed unlikely that plants operating in fields at ambient temperatures could achieve what chemists could accomplish in the laboratory only with considerable difficulty.

The proof that plant systems might be able to fix nitrogen was a far from trivial undertaking. At first sight, one would only have to repeat the van Helmont experiment of 200 years earlier. However, nitrogen is an integral constituent of all living things. How could one detect the extra nitrogen fixed by plants when they were already full of nitrogen? In any case, would all plants fix nitrogen, and would they do so at every stage of their development? How could one grow plants in any kind of normal environment and ensure that they were not contaminated by new nitrogenous material brought in by insects, the wind, or even muddy boots? The conclusive experiment would require the most careful work, and the results would have to be relayed to a very sceptical audience. In addition to all that, the guano deposits of Chile and Peru were being exploited to feed the burgeoning populations of Western Europe and the United States, so there did not appear to be any commercial drive to find new methods of providing fertilisers for agriculture.

Boussingault and His Followers

One of the first researchers to tackle this problem was Jean Baptiste Boussingault.[24] He was born in Paris in 1802 and seems to have travelled widely in his youth, including in Colombia in South America, where he would doubtless have learned of the value of guano. He was certainly educated as a chemist, and he carried out some agricultural manuring experiments on his father-in-law's farm until about 1841. His field experiments persuaded him that the masses of carbon, oxygen, hydrogen, and nitrogen in a crop were greater than the masses of these elements provided by the manures. Coupled with crop rotations involving potatoes, clover, wheat, turnips, and oats, the annual analyses showed that the increase in nitrogen was somehow associated with the clover. The carbon could be clearly ascribed to the already well-known ability of plants to take up carbon dioxide during photosynthesis. The hydrogen and oxygen could be attributed to water. The obvious question was: From where did the nitrogen originate, the soil or the air?

In 1845, Boussingault was appointed to the Chair of Agriculture at the Conservatoire des Arts et Métiers in Paris. It should be remembered that at that time a chemist was just as likely to be involved in agriculture or medicine as in what we would today recognise as chemistry. Boussingault's experimental results, reported in 1858, seemed to show that peas also accumulated more nitrogen than expected, but, due to von Liebig's opposition, Boussingault repeated his growth experiments with peas housed in large glass containers fed with washed air mixed with carbon dioxide. The results were generally negative. By 1860, Boussingault concluded that his first experiments were false positives and that von Liebig was correct in that dinitrogen of the air was not the source of the increased nitrogen content. Boussingault was not the first scientist whose results were rejected by von Liebig and thus generally discounted but who was subsequently proved to be right.

However, not everyone was convinced of von Liebig's authority in this matter. Another Frenchman, named Georges Ville, also carried out growth experiments, using sealed bell jars, and claimed increases in the nitrogen contents in a range of plants, only some of which were legumes. There followed a whole series of researches by various workers that were not conclusive, some of which were positive and some negative. Even the possibility of oxidation of dinitrogen by various chemical and physical agents was advanced to explain the contradictory data. Definitive field experiments were lacking.

The pioneers of intensive long-term field experiments were Lawes, Gilbert, and initially, in part, Pugh, some of whose experiments are still proceeding today, 150 years after they commenced (see chapter 7 for more information on these long-term researches). Sir John Lawes was a landowner, Sir Joseph Gilbert was a chemist who had studied under von Liebig at Giessen, and both became interested in von Liebig's chemical approach to fertilisers and plant growth. Evan Pugh was an American of Welsh extraction who had also studied with Liebig before he collaborated with Lawes and Gilbert. He died in Pennsylvania in 1864 as the result of an accident, not yet aged thirty-seven, and married for less than three months. Lawes, Gilbert, and Pugh set up the famous experiments on the Broadbalk field at Rothamstead. They grew wheat continuously on the same plots, each treated in the same way every year. Some received no fertiliser at all, some received farmyard manure, and some received different mixtures of artificial fertilisers. These experiments showed that the crop yields improved dramatically when a form of nitrogen was added to a field and that otherwise the yields tended to drop, though relatively slowly. Evidently, the test fields contained considerable resources of fixed nitrogen that were mobilised relatively slowly. Further experiments with rotations involving legumes showed that these were able to use sources of nitrogen that were apparently not available to non-legumes. Clover was found to be particularly efficient at mobilising nitrogen and promoting higher crop yields, something that generations of farmers, since at least the time of the Romans and the early Chinese, already knew from their direct experience. The source of this nitrogen, whether the soil or the air, was the question that these results posed.

These field experiments were always open to the criticism that the observations were caused by contamination. Von Liebig was certainly not convinced of their reliability. The only way to be sure would be to grow plants in enclosed systems, and that Gilbert and Lawes proceeded to do using some of the apparatus sent from France by Ville. They grew cereals and legumes in glass bell jars sealed at the base with mercury to prevent the access of solid or liquid contaminants and kept under a slight positive pressure of purified air to prevent gaseous impurities from entering. As discussed earlier, von Liebig had already demonstrated, at least to his own satisfaction, that ordinary air was a vehicle for supplying ammonia to plants. As far as possible, Gilbert and Lawes eliminated every source of nitrogenous and other contamination, even oxidants such as ozone, O_3, which is capable of oxidizing dinitrogen of the air to oxides of nitrogen. Ordinary dioxygen, O_2, is not able to do this. The results were convincing—they obtained no firm evidence for any increase in the nitrogen content of the systems. What they had actually proved, and what is still accepted today, was that plants cannot fix atmospheric nitrogen. Boussingault was convinced of this, and he told Gilbert so. Von Liebig was doubtless also very happy with these results. Nevertheless, field experiments at Rothamstead and elsewhere continued to suggest that legumes, and particularly clover, somehow seemed to be able to mobilise nitrogen, rather like the lager of some modern advertisements (legumes, the plants that can use nitrogen that other plants cannot reach). And in truth, they really could.

Hellriegel and Wilfarth

The breakthrough came with the work of Hermann Hellriegel and Hermann Wilfarth from the Bernberg Experimental Station in Prussia. It is remarkable that Prussia and Germany featured in so many of the scientific advances of the later nineteenth and earlier twentieth centuries. The reason may be as much political as scientific. There was a strong movement to foster German unity that came to fruition in 1870, though unification was by the sword of Prussia rather than by the pens of liberal politicians. Once the political unity of Germany had been forged, certain circles began to ask how it could begin to rival world powers such as Britain and France. Certainly, building an empire in the British or French mould was hardly feasible for Germany. The British and French would not countenance it. However, many people, including the extraordinarily influential von Liebig, saw science as a basis for challenging the industrial might of Britain and France. The consequence was a considerable emphasis on science education and on the growth of science institutes, encouraged and financed by the government and the Emperor (Kaiser) and aimed at advancing industry.

Even Roger Bacon in the thirteenth century had suggested that rulers who wished to see further than their enemies (using lenses) and to destroy them more effectively with explosives (using gunpowder) should invest in scientific research, so this was hardly a new idea. So while Britain was prepared to leave such matters to individual entrepreneurs and to wealthy gentlemen with a scientific interest such as Lawes, in Prussia, and later in Germany, the government invested in science. In Britain, W. H. Perkin invented coal-tar dyes, rapidly made a fortune, and retired to lead the life of a gentleman undertaking genteel and rather ineffective research funded from his own resources. In Germany, coal-tar dyes became an industry that supplied the world.

The observations that finally solved the nitrogen conundrum were really very old. The roots of legumes often carry excrescences that can reach a considerable size. They were illustrated in exquisite detail as early as the sixteenth century and are even depicted in a drawing dating from 1541 (see also figure 1.6). What these organs were and what they did were completely unknown. In fact, we now know that certain other families of plants also carry these growths, though it is still a rare phenomenon other than with legumes, of which they are, however, characteristic. Today, these growths are called nodules, and they are the result of an infection that enters the plant via the legume root hairs or through cracks in the epidermis of the roots. Boussingault had already noted that his best leguminous plants seemed to be those with nodules on the roots, and he suspected that they might have something to do with nitrogen fixation. Other people made similar suggestions.

It is tempting to suggest that the reason that Lawes et al. were unable to detect "new" nitrogen in their systems was that they had been too careful. Using sterilised soils meant that all infectious agents were destroyed, and it was just such an infectious agent that was the basis of biological nitrogen fixation. It was the particular genius of Hellriegel and Wilfarth to realise that what was needed to fix nitrogen was both a legume and an infectious agent.

By growing cereals and legumes in pots, Hellriegel and Wilfarth showed that whereas cereals responded systematically, consistently, and predictably to added fixed nitrogen, legumes did not. There seemed to be some factor that was not being controlled that affected the growth of the legumes but not the cereals. By 1885, it was suspected that the nodules on the roots, which were by then known to contain bacteria-like organisms now termed bacteroids, might somehow be the cause of the discrepancy, perhaps because they scavenged nitrogen from the soil. In 1886, Hellriegel and Wilfarth performed a key experiment, growing pea plants in a sterile medium without fertilizer nitrogen and, in parallel, pea plants under the same conditions to which they added either a suspension of raw soil or a similar suspension after heat sterilisation. The limited results were quite clear: Something in the unsterilised soil had enabled the peas treated with it to grow more efficiently. Later experiments confirmed that cereals did not respond to the raw soil suspension, though legumes always did. Hellriegel and Wilfarth noted that the well-grown legumes were also nodulated on their roots. Most of these experiments were interpreted in terms of the dry weight of the plants produced rather than in terms of nitrogen content. That would come later.

The conclusive experiments were carried out with plants grown in glass vessels, and everything used was sterilised. The pots were set up on trolleys so that the plants could be grown in the air and sun but moved under glass cover if rain threatened. Birds were excluded. The account of the research of Hellriegel and Wilfarth that was presented by Nutman[24] to commemorate the centenary of their proof of the reality of biological nitrogen fixation in 1887 is particularly readable and informative and contains copious references to the original literature. They determined plant yields (dry weights) and nitrogen contents for various cereals and legumes treated in a variety of systematic different ways and concluded unequivocally that, without fertiliser nitrogen, only nodulated legumes showed vigorous growth, though such nodulation did not guarantee it.

Hellriegel and Wilfarth also carried out the ingenious experiment sketched in figure 4.7. By dividing the plant root system into two parts and inoculating only one-half with soil while the other part remained sterile, they showed that the nodules were indeed produced from something (presumably a bacterium of some kind) in the soil. By the end of 1887, it was clear that the root nodules were the seats of nitrogen fixation. Biological nitrogen fixation was a reality, and this explained what farmers had already known for at least 2000 years: Legumes could help to restore soil fertility. What the farmers could not have known was that legumes do this by fixing nitrogen and leaving it in the soil when they die and decay.

Hellriegel and Wilfarth also showed that different soils contain different amounts of the nodule-inducing agent and that there was some specific relationship between the agent and the legume. Agents for one soil source seemed

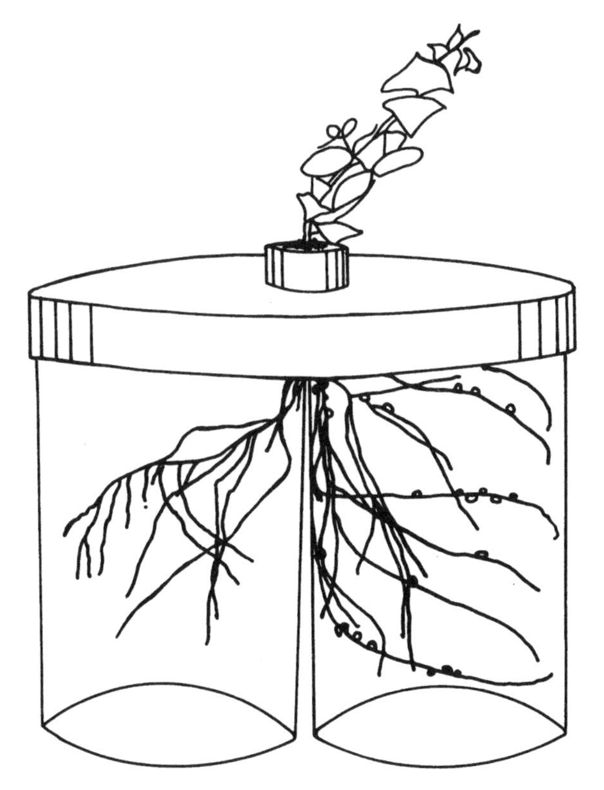

Figure 4.7. The crucial experiment of Hellriegel and Wilfarth to prove that the nodulation of the pea plant was caused by some kind of infectious agent in the soil. Only the half-root inoculated with the raw soil suspension bore nodules (reproduced with permission of the Royal Society and the author from P. S. Nutman, *Philos. Trans. R. Soc. London*, B 317, 11 (1987).

to infect specific types of legumes with greater efficiency than they did others. They knew that the agent could be destroyed by heat and desiccation, and that clearly suggested that it was alive.

The Chemistry of Biological Nitrogen Fixation

The agents responsible for the formation of the root nodules were isolated in 1888. They are now termed rhizobia, as mentioned in chapter 1, and, left to themselves, they live normal happy bacterial lives in the soil, where, however, they cannot fix nitrogen. There are many varieties of rhizobia. Once they invade the root hair of an appropriate legume, they take up the form termed a bacteroid, and bacteroids can reproduce and fix nitrogen (figure 4.8). They are suspended within the nodule and within a membrane, in a solution

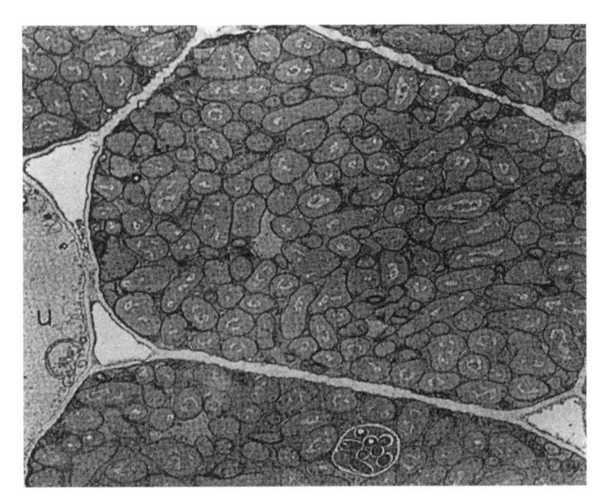

Figure 4.8. Mature infected cells of a buried clover nodule filled with bacteroids. The cell labelled "u" has not been invaded, whereas the lower cell, only part of which is shown, also displays a cross section of an infection thread filled with bacteria yet to be changed into bacteroids (reproduced with permission from P. J. Dart and F. V. Mercer, *Archiv. Mikrobiol.*, 49, 209 (1964) © Springer Verlag.

of a red material that seems to be chemically similar to mammalian haemoglobin. It is called leghaemoglobin and is believed to perform a function related to that of mammalian haemoglobin. It removes dioxygen from the air that diffuses into the nodule and leaves the dinitrogen. It is thus performing a protective function, for the bacteroids are dioxygen-sensitive and would otherwise be harmed. However, leghaemoglobin also transports dioxygen to the bacterial cell mitochondria, where it is used in the sequence of reactions that produce ATP (adenosine triphosphate). The bacteroids have a symbiotic relationship with the host plant. The plant provides the bacteroid with energy in the form of sugars such as glucose, and in return the plant receives fixed nitrogen in the form of ammonia.

The overall chemical equation for the biological fixation of nitrogen under the most efficient conditions is believed to be usually that shown below, where P_i is used to denote inorganic phosphate. This is clearly an energy-intensive process, with two molecules of ATP hydrolysed to ADP (adenosine diphosphate) for every electron (e) generated and used:

$$N_2 + 8H^+ + 8e + 16MgATP \rightarrow 2NH_3 + H_2 + 16MgADP + 16P_i$$

The actual transformations depicted are very much more complex than this single equation would imply and involve many related steps. The whole process is mediated by the enzyme complex called nitrogenase. The equation above shows that one molecule of dinitrogen together with some hydrogen ions from water are converted to two molecules of ammonia, a basic building block for amino acids. One molecule of dihydrogen (H_2) is also evolved, and some tentative attempts have been made to collect it from growing legumes to use as a fuel. At some time in the process, the biological energy carrier ATP (in the form of its magnesium salt, MgATP) is hydrolysed to MgADP, and phosphate (PO_4^{3-}) is liberated. To a chemist, the equation above simply describes this complete transformation in quantitative terms.

Plants use the energy of daylight and the process known as photosynthesis to form ATP, generating dioxygen at the same time:

$$6H_2O + 6CO_2 + \text{light energy} \rightarrow C_6H_{12}O_6 + 6O_2$$

$C_6H_{12}O_6$ is the chemical formula for glucose, and since the nitrogen-fixing mechanisms of most organisms are dioxygen-sensitive, some stratagem may be necessary to protect oxygen-sensitive operations from oxygen damage. This is not a problem in leguminous systems, which photosynthesise in the leaves but fix nitrogen in the nodules, but cyanobacteria (see below) have developed specialised cells to overcome the difficulty. These specialised cells fix nitrogen but do not photosynthesise.

ATP hydrolysis is a general biological method for the supply of energy for physiological processes within a cell. The legume uses the glucose to produce ATP, and the hydrolysis of the ATP to ADP by the rhizobia in a legume symbiosis releases phosphate (PO_4^{3-}) and energy. It is this energy that is used to split the nitrogen-nitrogen bond of the N_2, leading eventually to ammonia.

The Occurrence of Biological Nitrogen Fixation

The symbiosis of rhizobia and legumes is widely exploited agriculturally and seems to be widespread in nature. As noted by Postgate,[25] more than 12,000 species of *Leguminosae* are known, and the majority can take part in nitrogen fixation, though fewer than fifty *Leguminosae* are used agriculturally. Plants that may be nodulated by rhizobia but that are not members of the *Leguminosae* family are rarer. Today, *Leguminosae* are defined in botanical terms as plants that bear fruit in pods that split open into two parts when they dry and scatter the seed. This definition does not imply that all *Leguminosae* are herbaceous plants, and indeed they are not. A particular example is the acacia, a tree. Further, this definition does not immediately explain why it is predominantly members of the family *Leguminosae* that are involved with symbiotic nitrogen fixation. The botanical definition of *Leguminosae* seems in some ways to be as superficial as the original Roman definition—that they bear fruit that may be gathered by hand.

There are other symbiotic systems that facilitate nitrogen fixation and that do not involve *Leguminosae*. For example, the alder tree and the bog myrtle form nodulated roots that do not contain rhizobia, and these are two of about 250 examples. The nodules are formed by filamentous bacteria called

Actinomycetes, and they are given the generic name of *Frankia*. The nature of the symbiosis is not as well-understood as that of the rhizobia. The agricultural significance of such symbioses is minimal, whereas the total fixation of nitrogen by biological systems is now reckoned to be of the order of 100 million tons per year worldwide.

There is a further group of nitrogen-fixing bacteria that may be involved in symbioses. They used to be called blue-green algae and are now more often termed cyanobacteria. These are the organisms sometimes responsible for the rapid growth in nitrogen- or phosphorus-enriched environments that precedes what is popularly termed eutrophication. Cyanobacteria form colonies of cells in chains, and certain members of the chain are specialised. Most of the individual cells photosynthesise, a process that here produces dioxygen. This would destroy nitrogenase enzymes present in the same cell, so nitrogen fixation takes place in other specialised cells called heterocysts, and these do not photosynthesise. In this way, the nitrogen-fixing apparatus is not exposed to oxidizing conditions, but there must be a means of communication between the individual cells that make up the chain, allowing the exchange of nutrients and ammonia between them. In other kinds of organisms, the separation of nitrogen fixation and photosynthesis is achieved by carrying out only one of these functions at any given time, perhaps as a function of the presence or absence of the daylight necessary for photosynthesis.

Cyanobacteria of various kinds can also live in association with plants, as they do with fungi to form lichens. Other cyanobacteria live on the leaves of ferns such as *Azolla*. The Chinese discovered long ago that the leaves of the tiny fern *Azolla* were good green fertilisers for rice paddies and for many centuries have grown these ferns and ploughed them in to nourish their crop plants. Even the giant fern *Gunnera* forms a symbiotic relationship with a cyanobacterium.

There is evidence that some termites have developed a symbiotic association with some types of bacteria. Termites seem to live in particularly low-nitrogen environments. However, such arrangements are not usual. There are many kinds of bacteria that fix nitrogen living free in the soil. Particularly well-researched are the Azotobacters and the Clostridia. The Azotobacters are aerobic (that is, they require dioxygen for respiration), and since they are unicellular organisms, they must have developed a mechanism to protect the nitrogenase. In fact, such organisms often only produce nitrogenase and fix nitrogen when they exist under conditions of nitrogen deficiency, which suggests that nitrogen fixation is not something they gladly undertake. It would appear that even those bacteria that require dioxygen in order to function have to find some way to protect their nitrogen-fixing apparatus, which seems to be exceedingly sensitive.

On the other hand, the Clostridia are anaerobic. Since they only thrive in the absence of dioxygen, they have no need to protect their nitrogenase, but they are, of course, rather restricted in their choice of habitat. All in all, the biological nitrogen fixation accounts for about 30% of the annual total of nitrogen fixed naturally and industrially.

The recognition of the widespread occurrence of biological nitrogen fixation has arisen only since the pioneering work of Hellriegel and Wilfarth. Research to establish this has been intense but this kind of work is essentially old-fashioned, involving collecting and classifying. It does little to establish what is involved in the fixation process itself.

Apart from this kind of classification work, perhaps the major significant observation concerning biological nitrogen fixation during the first half of the twentieth century was the demonstration by H. Bortels in the 1930s that the presence of either molybdenum or vanadium enables bacteria to fix nitrogen.[26] This was achieved by growth experiments on various media and was essentially what would now also be regarded as old-fashioned science. The technique employed was to grow a legume or a bacterium, in this case *Azotobacter chroococcum*, on carefully purified growth media that had been deprived of various elements. This kind of experiment is not trivial to undertake. First, it is not easy to ensure that every last trace of a given element has been removed from the experimental system, and we now know that some organisms, amongst them *Azotobacter chroococcum*, are exceedingly efficient at scavenging trace metals from the environment. Second, though it is possible in this way to prove that certain elements are necessary for the life of the organism under study, it is not so easy to show what part that element plays in the organism's metabolism. For example, iron is necessary for so many different functions in most organisms that growth experiments involving simple deprivation of iron cannot tell us very much about what iron does. For this reason, the sometimes contradictory results of Bortels and others were not generally given much emphasis. This kind of work was pursued extensively until about 1960, when the whole fixation field suddenly burst into new life with the first isolation of cell-free extracts of the enzyme responsible for biological nitrogen fixation, nitrogenase.

Nitrogen Fixation at the End of the Nineteenth Century

Clearly, von Liebig and the generation of chemists who doubted that plants could fix nitrogen were both right and wrong. They were right in believing that plants cannot fix nitrogen. However, there are agents in the soil that can infect the roots, and within those roots, these agents can fix nitrogen. Maybe they can fix nitrogen in the soil, but that wasn't known or perhaps not even open to investigation in 1887. Chemists were wrong in believing that biological nitrogen fixation is not a fact, but the establishment of this fact raised even more puzzling problems. How can bacterial agents fix nitrogen when chemists find it so difficult? If it is not credible that plants can do it, that bacteria might is even more incredible. How might they do it? What do they know that von Liebig and all his successors do not? These questions are still not fully answered nearly 120 years later.

In 1898, Sir William Crookes made his well-remembered address to the British Association, calling on the chemists to save the world (or what he regarded as the civilised, white part of it) from starvation. The deposits of nitrate in Chile and Peru were nearly exhausted, and the challenge was stark. At the same time, biology also seemed to offer a partial way out. Legumes, which the Romans had defined simply as plants whose fruits could be gathered by hand, were revealed as plants capable of fixing nitrogen, albeit with a little help from some other organisms. However, the development of science at the end of the nineteenth century was such that it would have been impossible then to learn very much more of the intimate details of biological nitrogen fixation. What was happening in the organism at the atomic or molecular level rather than at the whole-cell level could not be determined with the knowledge then extant. It wasn't possible to advance understanding of nitrogen fixation at the chemical and molecular level until the necessary techniques and theoretical framework had been developed. This was to take another fifty years.

Further advances in biological understanding might not have been feasible at that time, but the same was not as true of chemistry. The need to obtain sufficient nitrate to ensure enough food to nourish your population and enough explosives to destroy your enemies lent an enormous impetus to the search for a reliable and efficient chemical method to fix nitrogen on an industrial scale. Chemical science was about to experience the fruits of the explosive growth in the basic understanding of chemistry during the previous half-century. Though Crookes did not know it, in 1898 the solution to the problem he had set for the world's chemists was already being developed.

CHAPTER 5

The Triumph of Industrial Chemistry

The Industrial Response to Sir William Crookes

In 1905, Sir William Crookes published a book entitled *The Wheat Problem* in which he reiterated what he had said in his British Association address of 1898. The content and tone are familiar[1]: "The fixation of nitrogen is vital to the progress of civilized humanity, and unless we can class it among the certainties to come, the great Caucasian race will cease to be foremost in the world, and will be squeezed out of existence by races to whom wheaten bread is not the staff of life." A whole gamut of processes for fixing nitrogen was described in a book published in 1914,[2] and in 1919 an eminent U.S. electrochemist, H. J. M. Creighton,[3] published a series of three papers entitled "How the Nitrogen Fixation Problem Has Been Solved." However, the broader story was only just beginning to unfold.

In about 1925, J. W. Mellor, in a justly celebrated sixteen-volume compendium,[4] simply took Creighton at his word and stated quite baldly: "The problem has since [Crookes' lecture] been solved." Mellor describes not one but six processes that he believed were of industrial significance. These were: (1) the direct oxidation of dinitrogen by dioxygen to yield, initially, nitrogen oxides, as was undertaken in the Norwegian arc process; (2) the absorption of dinitrogen by metal carbides, subsequently developed as the cyanamide process; (3) the reaction of dinitrogen and dihydrogen by what has become known as the Haber process, or, more justifiably, the Haber–Bosch process; (4) the reaction of dinitrogen with metals, followed by treatment of the resultant nitrides with water; (5) the reaction of dinitrogen with carbon to form cyanides; and (6) the oxidation of dinitrogen during the combustion of coal or natural gas. Of these, only the first three really reached the stage of industrial exploitation, and only the Haber–Bosch process has been applied to any degree of significance since about 1950. The history of these three major developments is traced below.

The Norwegian Arc Process

One of the first industrially significant reactions to be developed at the beginning of the twentieth century had already been known for more than 100 years. This was the reaction of dinitrogen and dioxygen to form nitrogen(II) oxide, nitrogen monoxide (sometimes still termed nitric oxide):

$$N_2 + O_2 \rightleftharpoons 2NO$$

This is an equilibrium reaction (meaning that it may proceed in either direction and that an undisturbed system will eventually reach a steady state in which the forward reaction exactly balances the back reaction). This reaction does not occur to any sensible extent under the usual atmospheric conditions; otherwise, the air we breathe would no longer be a mixture with dioxygen and dinitrogen as its two principal components but would also contain nitric acid, HNO_3. There is even a rather gruesome[5] science-fiction book, *The Nitrogen Fix*, by Hal Clement, which is based upon the proposition that this reaction had actually occurred, so humans were living in artificial environments protected from the all-encompassing acid. That aside, Mrs. B. and her students mentioned in chapter 4 well knew that lightning provokes the reaction of dioxygen and dinitrogen, and this forms part of the natural nitrogen cycle (chapter 1, figure 1.8). The reaction needs a lot of energy to make it proceed. Nitrogen monoxide, in contrast to its two relatively benign components, is a very reactive and unpleasant material. In the presence of dioxygen and water, ultimately it gives nitric acid. This process can be represented by the following equation:

$$4NO + 3O_2 + 2H_2O \rightarrow 4HNO_3$$

Though concentrated solutions of nitric acid are rather nasty, this very dilute atmospheric solution can give rise to nitrates in the environment, to be taken up by plants and used in plant metabolism.

It is often believed that the celebrated scientists whose names are most widely remembered make their discoveries in some flash of inspiration that vouchsafes them an understanding denied to their less gifted colleagues. In fact, this is rarely the case, and most breakthroughs are an incremental advance on what has been achieved by scientists from all over the world, from many different countries and often from different centuries. The story of the discovery of the various nitrogen-fixation processes[4] is a good illustration of this.

The Origin of the Norwegian Arc Process

In about 1775, Joseph Priestley passed electric sparks through air contained in a closed vessel. The volume of the gas in the container contracted, eventually to about 20% of the original. It seems certain that Priestley produced NO, then NO_2, and finally a solution containing nitric acid (HNO_3) and nitrous acid (HNO_2). He could not have understood all this detail at the time. This observation was extended by Henry Cavendish, who showed that dinitrogen was consumed as the nitric acid was produced. Later, Cavendish showed that the ultimate product of such sparking experiments in the presence of alkali was nitre, potassium nitrate. This method of fixing nitrogen was established before the end of the eighteenth century, though it was hardly an industrial process.

Throughout the nineteenth century, attempts were made to exploit this dinitrogen/dioxygen reaction. It was evident that very high temperatures would be required, and at that time these could only be achieved by using electric sparks. Initially, these were not sustainable either in amount or quantity to enable a reliable industrial process to be developed. By about the middle of the nineteenth century, continuous electric discharges were available, and it was realised that these also could provoke the formation of nitrogen monoxide and thence nitric acid. British Patent[6] 1045 to fix nitrogen in this way was taken out on behalf of Madame L. J. P. B. Lefebure in 1859. She also took out a similar French patent in the same year. Madame Lefebure was possibly the first female chemical engineer and certainly one of the first. She proposed to make nitric acid by passing sparks through air contained in a large closed globe and absorbing the products in alkali. Neither of her patents was ever exploited industrially. Even Crookes himself worked in this field, and he confirmed that the flame sometimes observed in air subject to an alternating discharge was really due to dinitrogen actually burning. However, as early as 1892, he wrote[7]: "The reason why, once nitrogen is set on fire, the flame does not spread throughout the whole atmosphere . . . is that the ignition-point of nitrogen is higher than the temperature produced by its combustion, and therefore the flame is not hot enough to set fire to the adjacent gas." One way around this problem was suggested in British patent 12,401 of 1905. Oxides of nitrogen were to be produced in what was essentially the cylinder of a large diesel engine and then absorbed in water. Today, this would be regarded as an environmental hazard rather than a commercial opportunity.

The use of single isolated sparks in an appropriate gas mixture was never likely to be an efficient way to fix nitrogen, but a continuous discharge in an electric arc was a different matter. The technology of electric arcs improved toward the end of the nineteenth century, and work on the first factory to produce acid by the arc process was started in Norway in 1903.[8]

The way this happened is, in retrospect, rather unexpected. On February 6, 1903, Professor Kristian Birkeland gave a lecture-demonstration at the

Kristiana University in Oslo. He had invented an electric or magnetic cannon or torpedo launcher that he attempted to sell to various governments over a number of years (figure 5.1). His idea was that since a conventional gun or cannon worked because the explosion of the propellant in the gun chamber produced very rapidly a large volume of gas, the expansion of which could be used to propel a missile out of the gun barrel, if one were to cause a similar rapid expansion of gas in a closed chamber by passing an electric spark, then this could also be used to power a gun shell. He needed an enormous current for the rapid generation of the very high temperature necessary for this to work properly. The lecture-demonstration was a spectacular failure due to a short circuit, but this did not discourage Birkeland, who was obsessive about his ideas.

Apparently, Sam Eyde, then a structural engineer, and Director Fredrik Hiorth had been wondering how to use the abundant waterpower of Norway for commercial purposes. On February 13, 1903, both Birkeland and Eyde dined together with a common friend, Gunnar Knudsen. There is some disagreement about who persuaded whom to do what. According to the official history of Norsk Hydro, Eyde, the engineer, realised that Birkeland, the physicist, had really developed an electric furnace to power his electric cannon and that such a furnace could be used directly to promote the sustained reaction of dinitrogen with dioxygen. Thus was the idea of the famous Norwegian arc process born. It required enormous resources of electric power, a method of producing a stable electric arc at a temperature of about 2000 °C, and sufficient engineering skill and resources to bring the whole process to maturity. Probably it was only in Norway at that time that all this could have been achieved. In any case, Birkeland and Eyde agreed to take out a patent concerning use of an electric arc to make nitrogen compounds from nitrogen

Figure 5.1. Professor Birkeland's electric cannon, from a picture in note 8. It is not evident how this operated, and this model was probably not meant to fire shells. In any case, the demonstration backfired spectacularly, which meant ultimately that the idea was exploited more fruitfully than Birkeland had envisaged (© Norsk Hydro ASA, reproduced with permission).

gas, and this they first applied for on February 20, 1903, only a week after their dinner. All they needed was a sufficient supply of power and some money. Eyde set out to provide them.

The Cyanamide Process

A second industrial process that was developed during the latter half of the nineteenth century was the cyanamide process. This is almost as venerable as the Norwegian arc process in its origins. The fundamental conversion in the cyanamide process is essentially the reaction of atmospheric nitrogen with a carbide to form a variety of cyanide called cyanamide. It can be represented by the following equation in which the metal involved is calcium, for reasons that will become evident below, but it is certainly not restricted to this element alone. This equation represents the reaction of calcium carbide with dinitrogen to form calcium cyanamide and release some of the carbon:

$$CaC_2 + 2N_2 \rightarrow 2CaNCN + C$$

The reaction occurs only at very high temperatures, so that, as with the arc process, an electric furnace seemed almost mandatory. Calcium carbide itself was made by fusing carbon in the form of charcoal with calcium oxide, otherwise called lime or quicklime, itself derived by heating calcium carbonate, which was obtainable as plentiful chalk or limestone. Calcium carbide generates acetylene when water is added to it, and acetylene is a flammable gas. The controlled addition of water to calcium carbide was used as the basis of the old-fashioned acetylene lamps, so at the turn of the nineteenth century and earlier calcium carbide was a material of general commerce.

However, there were earlier observations that also led to the development of the cyanamide process.[3] Scheele probably observed that dinitrogen reacts with heated sodium carbonate and carbon as early as 1775, and in 1839 iron was observed to catalyse this reaction, which generates cyanides. One John Swindells took out British Patent 8036 in 1839 describing the manufacture of cyanides from a mixture of sodium or potassium sulfate, calcium oxide or carbonate, iron filings, and coal heated in a furnace in air. Associated products were carbonates and sulfides.

There are reports from 1813, 1819, and 1826 of lumps of fused salts that were to be found in the bottom of iron blast furnaces. These were assumed to be chlorides or carbonates, and it was only in 1835 that it was recognised that this kind of material also contained cyanide.[9] The phenomenon became more and more common as improved technology allowed the temperatures of operation of blast furnaces to increase. The combination of an extremely high temperature, a metal such as iron, carbon in the form of coal, and atmospheric nitrogen is just what one might now expect to produce cyanide. When new hot blast furnaces were introduced on the Clyde in Scotland in 1837,[9] the workmen soon noticed "peculiar exudations of fused salt" forming upon the walls of the furnace. One particularly observant workman apparently noted that this material was alkaline and took it home to assist his wife in washing the family clothes, where it was apparently used as a substitute for soap. Chemical analysis later showed the material to be a mixture of potassium carbonate, $K_2(CO_3)$ (45.8%), and potassium cyanide, KCN (43.4%), together with other, possibly noxious materials.[10] Apparently, the furnace produced cyanide for at least three more years, but it is not recorded how long the workman's wife survived to wash clothes, though Breneman says somewhat laconically[9] that she used the material "for a time". A German commentator of the period noted drily that the material was not suitable for domestic use.

Throughout the nineteenth century, there were reports of cyanide (and a related material, cyanate) being produced in blast furnaces. The production seems to have been commonly recognised, but it is now not clear whether this was regarded as a hazard, an advantage, or merely a nuisance. Breneman[9] records having observed cyanide formation in blast furnaces on at least two occasions. Bunsen and Playfair reported in 1845[11] that a furnace at Alfreton, Derbyshire, in the north of England, yielded 224.7 lb (there were no SI units then, though in a footnote they express a preference for the "French" units of metres and grammes) of cyanide in 24 hours. It seems to have been produced in a very restricted part of the furnace, so temperature and gas composition were probably critical. The conclusion, after considerable discussion, was that hot carbon and atmospheric nitrogen could react to give cyanide and that, as potash was generally present, the final product was potassium cyanide. Water also played a critical role. If it was present, then cyanide was still a product, but the main compound then produced was cyanogen, which has the formula C_2N_2:

$$2C + N_2 \rightarrow C_2N_2$$

Either way, it was quite clear by 1850 that industrial fixing of nitrogen by this kind of route was feasible. However, the product, cyanide, was hardly going to be of direct use as a fertiliser, though one might have envisaged applications in dyeing (and, less benevolently, in dying!). An Anglo–French attempt to commercialise cyanide preparation seems to have started near Paris in 1843.[9] In this endeavour, the production of potassium ferrocyanide (then termed yellow prussiate) reached about fifteen tons per year, but the initial compound formed was actually potassium cyanide. The process used dried charcoal that had previously been soaked in a solution of potassium carbonate, and this was heated white hot in clay retorts through which the exhaust gases from coal fires were passed. This was presumably to heat the dinitrogen of the air and remove dioxygen before introducing the gases into the furnace.

Since Britain was, at that time, truly the workshop of the world, the process was moved from Paris to Newcastle-upon-Tyne in 1844. There it was set up on a larger scale, producing yellow prussiate at the rate of more than a ton per day. The process was carried out in cylinders made of firebrick, ten feet long, two feet in diameter, and with walls nine inches thick. Air was admitted to them through layers of charcoal. The cyanide-bearing charcoal was dropped into tanks of water containing a suspension of spathic iron ore (siderite, essentially iron carbonate), and the solution of yellow prussiate thus produced was then evaporated to induce crystallisation.

This complicated process was only operated for two or three years, and it was abandoned in 1847. The cost of the salt was said to be less than 2 francs (French) per kilo, which translated then as about $0.20 (U.S.) or £0.05 per kilo, but this was not a commercial price. Part of the reason must have been that the retorts suffered from excessive corrosion and needed replacement after several months. The toll on the operators must also have been considerable, though this was not a pressing consideration of the time. Throughout the rest of the nineteenth century, attempts were made to develop an industrial process for the manufacture of cyanide that would also be a commercial success, but Breneman[9] could write in 1889 that: "The history of later attempts to utilise the nitrogen of the air for the manufacture of cyanides, shows, up to the present time, no commercial success."

This did not prevent the ground being laid for the cyanamide process. In about 1860, it was found[12] that a material, supposedly barium cyanide, $Ba(CN)_2$, could be prepared by passing air over a hot mixture of barium oxide and carbon and that treatment of this cyanide with steam at 300 °C produced ammonia. With hindsight (see below), it seems probable that the key component of this material was not just barium cyanide but barium cyanamide, BaNCN. If this is indeed the case, one might represent the ammonia generation by the following equation:

$$BaNCN + 3H_2O \rightarrow BaCO_3 + 2NH_3$$

It would appear that people were ringing the changes on the alkali used, and barium was a logical replacement for potassium. Many patents, some in the United States, were taken out to protect variants of this process. Those that involved the formation of cyanides were of rather limited commercial application. However, clearly their decomposition by steam to yield ammonia was likely to be much more useful, particularly because the nitrate deposits of Chile were already realised to be inadequate to support the fertiliser requirements of Western Europe and the United States. Even Ludwig Mond, one of the founders of the modern British chemical industry in the form of what eventually became Imperial Chemical Industries (ICI), took out U.S. Patent 269,309 in 1882 to make cyanides and ammonia. None of the processes proved to be commercially viable.

Breneman's extensive review[9] was published in 1889. He considered all the factors that seemed to influence the reaction of carbon and dinitrogen, and his salient conclusions were as follows: "The manufacture of cyanides and ammonia from nitrogen of the air, while quite possible on the small scale, is beset with difficulties . . . which . . . have been insurmountable." Though he admitted that ordinary domestic coal fire might actually also be producing ammonia that went up the chimney, he was not very optimistic that cottage production of ammonia by the home fireside was very likely. He recognised the need for a strong base or alkali, and he also suggested that elevated pressure might increase yields. It is evident that at that time some people, at least in the United States, were still manufacturing cyanides from organic wastes, but the whole industry was about to be revolutionised by two developments. One was the increase of by-product isolation from coal-gas manufacture, and the other was the recognition that isolated carbides, and particularly calcium carbide, could be made to react directly with dinitrogen to give hitherto unrecognised materials and, in this particular case, calcium cyanamide.

Friedrich Wöhler first prepared calcium carbide in 1862. It required the reaction of calcium oxide (lime) and coke (impure charcoal) at temperatures above 2000 °C. Such temperatures were achieved in industry most easily and controllably using an electric furnace, and calcium carbide was produced industrially in this way by about 1890. Suitable electric furnaces could not have been built and used very much before then. The direct commercial use envisaged for calcium carbide was as a source of acetylene for lighting, but this was never as widespread as might have been expected. A modern furnace for the manufacture of calcium carbide is shown in figure 5.2.

The critical discovery in the fixation story then seems to have been by A. Frank and N. Caro, who knew by 1898[13] (the year of Crookes' historic

Figure 5.2. A modern calcium carbide furnace in Landeck, Austria. This furnace is considerably smaller and more efficient than the early furnaces, but the need for high temperatures is still evident from this picture. Reproduced by kind permission of Donau Chemie AG, Vienna, Austria.

address) that barium carbide, when heated with dinitrogen, forms a mixture of barium cyanide and barium cyanamide. It would appear from several of the patents that they filed that Frank and Caro's initial aim was to increase the yield of cyanide and reduce that of cyanamide. Later, it was found that relatively pure calcium cyanamide could be obtained from calcium carbide under similar conditions. Calcium cyanamide was first seen as an intermediate for the production of sodium cyanide, but in 1900 it was reported that it could be converted to ammonia by the action of superheated steam. The road to the use of calcium cyanamide directly as a fertiliser, and also to the industrial production of ammonia from it, was thus open.

The Origin of the Haber–Bosch Process

It is evident that the key to the successful production of fixed nitrogen, both by the Norwegian arc process and via cyanamide, was access to high temperatures generated by electricity. Before the technical developments that enabled suitable electric furnaces to be built, the chemistry of such nitrogen-fixation processes must have remained an academic curiosity. Nevertheless, the basis for the third industrial process listed above was being laid even as the two others were slowly coming to fruition. The fact that dinitrogen and dihydrogen can form ammonia under the appropriate conditions was well-known relatively early in the eighteenth century. Possibly, von Liebig knew about the chemistry underlying the arc process, cyanamide formation, and the dihydrogen/dinitrogen reaction, and it is likely that Davy also knew much of this chemistry about 100 years before any of these processes were developed. This only goes to confirm that no scientist ever makes a fundamental discovery that owes absolutely nothing to his predecessors. With this in mind, it is interesting that Breneman[9] in his 1889 review states, almost as a throw-away, that "Graham Young, in a process for the manufacture of ammonia patented in England in 1880, suggested the use of electricity to effect the union of nitrogen and hydrogen. The principle had also been mentioned in an English patent issued to Chisholm and Kent in 1860. No practical result has yet come from these suggestions." Another British patent of 1881 describes the manufacture of ammonia from dinitrogen of the air and dihydrogen obtained from steam and using catalysts either of iron or platinum. The air was freed from dioxygen by burning zinc metal in it! There is no clear evidence that any such processes were ever used commercially.

In 1903, Fritz Haber and his collaborator Alois Mittasch clearly demonstrated for the first time that it should be possible to develop an industrial process to produce ammonia from dinitrogen and dihydrogen. They laid the basis for the industrial development of the Haber–Bosch process, familiar to almost every tyro chemist. Although modern accounts of the work of Haber and his collaborators often give the impression that the process was sprung from nowhere upon an amazed and grateful scientific and industrial community, who immediately recognised its significance, it should already be obvious that things could not have been at all like that.

The Haber–Bosch process reaction is the simple union of dihydrogen and dinitrogen to yield ammonia:

$$N_2 + H_2 \rightleftharpoons NH_3$$

The double arrow in this case means that the reaction is reversible and that as well as dinitrogen and dihydrogen being capable of forming ammonia, ammonia itself can also decompose to give dinitrogen and dihydrogen. Under any given set of conditions in a closed system, both the forward and reverse reactions will proceed, and the result will eventually be a balance, an equilibrium. For this reaction to be useful, the position of this equilibrium, whether toward the right-hand side or the left-hand side of the equation, and the rate at which equilibrium can be achieved are critical factors.

The understanding that ammonia was constituted from nitrogen and hydrogen only clearly tempted early chemists to try a direct synthesis from the elements. The most obvious approach would be to mix dinitrogen and dihydrogen and to try to persuade them to react. The electric spark, as employed by Priestley with dinitrogen and dioxygen, would have quickly come to mind. In fact, Priestley in 1790 had observed the decomposition of ammonia into its elements when it is passed through a red-hot glass tube, and Davy claimed in 1807 that ammonia is formed during the electrolysis of distilled water in air. Though the synthesis and decomposition of ammonia by various routes were investigated throughout the nineteenth century, no obvious industrial process was evident in 1898. It had become apparent since the 1870s that various electric discharges could produce ammonia from its elements, but how much and how quickly were always in question. Some careful experimentation in 1884 by William Ramsay and S. Young showed that the decomposition of ammonia in a hot tube is temperature-dependent. Because all their decomposition experiments always yielded a small amount of apparently residual ammonia, they also wondered whether the formation of ammonia from its elements took place under the same conditions. They used iron filings as a catalyst, but the results were not unequivocal. Nevertheless, they must have suspected that some kind of equilibrium exists, for earlier studies on both the synthesis and decomposition of ammonia never seem to have given 100% yields of the desired products.[14]

It was probably evident by this time that the two basic questions to be answered before an industrial process would become feasible were: (a) Would the amount of ammonia produced be large enough to be exploited commercially? (i.e., is the position of the equilibrium in the synthesis reaction favourable enough?); and (b) How fast is the reaction and is there a suitable catalyst? (i.e., can equilibrium be reached quickly enough?). This is where Fritz Haber entered the picture.

Fritz Haber received the Nobel Prize for chemistry in 1918 for his "discovery" of the process that bears his name. His Nobel lecture, delivered in June 1920, after the end of World War I, summarises the position that existed when he became interested in the nitrogen problem. The reaction of dinitrogen with dioxygen to form nitrogen(II) oxide (nitric oxide or nitrogen monoxide, NO) was being intensively investigated and, like the dinitrogen/dihydrogen reaction, is also an equilibrium reaction. However, as with the vast majority of chemical reactions, the reaction rate is also lower at lower temperatures, so some kind of compromise (a higher theoretical yield but achieved more slowly) between these two opposed limitations might have been possible. The dinitrogen/dioxygen process had seemed to be possibly viable commercially in 1907, and Haber, in his Nobel address,[15] stated that with ". . . a number of excellent assistants I therefore studied for some long time the synthesis of nitric oxide by electrical discharge." Haber used lower temperatures and pressures than had been employed hitherto, and he certainly obtained some improvement in yield. He also considered the use of explosive reactions, such as that of acetylene with air, during which nitrogen(II) oxide is indeed produced in considerable yields, but his final conclusion was that ". . . the technical solution was to be sought in the direct combination of [di]hydrogen with [di]nitrogen." It is evident that Haber had, from the beginning, a clear idea of what a successful industrial process, as distinct from a laboratory process, might require.[16] In this he differed from many of his contemporaries. He was Professor of Physical Chemistry and Electrochemistry and Director of an Institute in the University of Karlsruhe. He had previously been an assistant of the Professor of Chemical Technology. Very few comparable positions would have been in existence at that time outside Germany.

There were many superficial objections to an industrial process involving dihydrogen. First, at that time there was no large-scale source of dihydrogen, let alone a technique for handling the vast quantities necessary. Second, it would need the separation of dinitrogen from the dioxygen of the air, no trivial process, whereas a dinitrogen/dioxygen reaction needed only air, water, and a sufficient temperature. Third, there was no evident catalyst, and finally, and crucially, Walther Nernst had studied the dihydrogen/dinitrogen equilibrium and come to the conclusion that it was probably not advantageous (the equilibrium was too far to the left in the equation) for exploitation as a synthetic process.

Haber went back to the rather ambiguous work of Ramsay and Young, reported in 1884, and some even earlier work reported by Deville in the 1860s. An assistant, G. Van Oordt, undertook the new research, starting in about 1904 (figure 5.3). Haber and Van Oordt circulated the reactant gases over an iron catalyst contained in a hot tube and then washed out the ammonia (with water?) at normal temperatures. This work confirmed that it was possible to obtain some ammonia (a fraction of 1%) using an iron catalyst and a temperature of 1000 °C, whether one decomposed ammonia or started with the

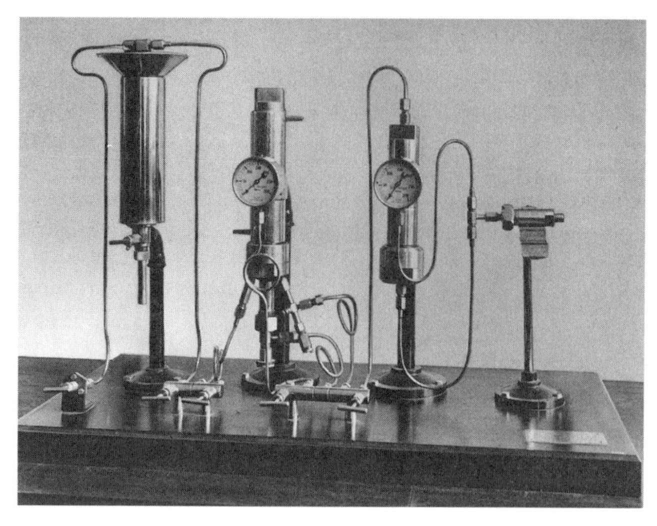

Figure 5.3. Original equipment of Haber. Apparently this was used in the demonstration with Mittasch in Karlsruhe in 1909 in which a yield of 80 g ammonia was achieved in one hour using an osmium catalyst. It was probably developed as early as 1905. Note that this is primarily a straight-through arrangement, and there is no obvious facility for recovering the ammonia and then recirculating the unreacted gases (courtesy Deutches Museum, München, Germany).

elemental gases. There certainly was an equilibrium involved. However, the value of the equilibrium constant (which measures the position of that equilibrium) seemed to show that the best results should be obtained with a 3:1 dihydrogen:dinitrogen mixture at a temperature below about 300 °C. Even increasing the pressure would not help much. At such low temperatures, the yield of ammonia would never be sufficient. By 1905, Haber decided that the outlook was too unpromising to justify pursuing the work further.[15]

However, the pressure to find a commercially viable route to fix nitrogen was recognised throughout the (wheat eating!) world. Sir William Crookes has already been cited. In 1906, the celebrated British chemist W. H. Perkin, the discoverer of the first coal-tar dye, visited the United States, and one of his pleasant duties there was to receive the first award of the Perkin medal, established by the American Section of the Society of Chemical Industry in Perkin's honour, to reward distinction in organic chemistry. The medal was presented on this inaugural occasion by Dr. William Nichols, President of the U.S. General Chemical Company. After describing Perkin's contribution to the general well-being of the world, and suggesting that Perkin could die a happy and satisfied man, now, said Nichols, it was the duty of chemists to save the world from starvation.[17]

Were it not for one happening that he could not quite accept, it seems likely that Haber would have withdrawn from the struggle to fix nitrogen commercially, satisfied that there was, as yet, no suitable process in the offing. This happening was that Nernst had investigated the dihydrogen/dinitrogen equilibrium, and his data became available in 1906. The results were not consistent with data that Haber himself had obtained earlier. There began a considerable dispute in which Nernst was eventually proved wrong. Haber decided to recheck the equilibrium data more carefully along with an assistant, Robert Le Rossignol, who, despite his name, was British. At this time, Haber was also already working on the dinitrogen/dioxygen reaction.

By 1908, Haber was convinced that the equilibrium might not be as disadvantageous as Nernst's data had originally implied. Haber and Le Rossignol concluded that what was needed for success was increased pressure (as implied by the Principle of Le Chatelier) and temperatures of 700 °C or higher. In his Nobel lecture, Haber remarks as an afterthought that Le Chatelier had also applied himself to the ammonia-synthesis problem but ". . . failure of the first attempts . . . led him to abandon the matter and to publish his deliberations only in the obscurity of a French patent taken out under a foreign name [*sic*]. This only came to my notice a long time after the successful conclusion of my experiments." The patent of Le Chatelier, taken out in 1901, was allowed to lapse, in part probably because the test apparatus blew up under the pressure applied to it, and the experimenters, though they might have lost eyes and limbs, apparently certainly lost heart. Clearly, this statement was also a damage-limitation exercise on Haber's part!

Research then quickly established that ammonia synthesis was indeed feasible at a pressure of 200 atmospheres in the system portrayed in figure 5.3, originally used in 1905. This was the maximum that Haber could achieve on a laboratory scale. The difficulties of circulating a gas at red heat and at elevated pressures using the technical means then available were considerable. By a judicious choice of catalyst, Haber and Le Rossignol were able to reduce the operating temperature to below 600 °C, which is perhaps 200 °C higher than the temperature generally used today. Apparently, uranium and osmium compounds gave rise to the best catalysts, and these are certainly no longer used. The system then produced a gas containing about 5% ammonia. A more understandable yield is expressed as 250 grammes per hour per litre of contact volume. In another enlightened insight, Haber realised that the synthesis reaction produced heat that could be used to raise the temperature of the incoming gases from room temperature to the desired reaction temperature, thus saving energy. With Le Rossignol, Haber developed an improved recirculating system so that the expensive and unreacted dihydrogen could be recovered and passed again over the catalyst, and the results obtained were convincing enough for Badische Anilin- und Soda-Fabrik (BASF) to take the project under its wing.

Up to that time, BASF had been investigating the possibility of obtaining ammonia by heating selected metals to high temperatures under dinitrogen to form nitrides and then treating the products with water to release ammonia. In addition, as we shall see, BASF also had an interest in the Norwegian arc process and had their own variant of it, called the Schönherr process, which was, of course, an oxidative method. Despite these other interests, by 1910 BASF was already constructing a factory at Oppau for the production of ammonia by what is now termed the Haber–Bosch process, and the second (and often ignored) genius involved in this particular story had entered the scene.

The development of the industrial process was a heroic achievement, but the practical details were also highly confidential at the time. Any chemist worth his salt already knew by 1913, when Haber and Le Rossignol described their work in detail in the open scientific press,[18] that the dihydrogen/dinitrogen reaction could be used on an industrial scale to produce ammonia. For two or three years before that, both Haber and BASF had been taking out patents concerning the synthesis of ammonia from dinitrogen and dihydrogen, so the chemical industry must also have been aware that something was up. What was not widely known outside BASF was the detail of how this was done commercially. The attempts made by others, and by the British in particular, to discover the secrets are, with hindsight, extremely funny but also rather distasteful. However, it is now necessary to describe the developments stemming from Norway during the period up to the beginning of World War I in 1914.

The Rush to Fix Nitrogen Industrially

The Norwegian Arc Process

Although the first steps toward the building of a factory to produce nitrogen(II) oxide (nitric oxide) by the arc[8] method were taken in 1903, there was an even more serendipitous factor at work than the chance meeting of Birkeland and Eyde at a dinner party with a mutual friend. In his memoirs, Eyde mentioned that he was aware of the "conférence sensationelle faite, en 1898, par le fameux savant Sir William Crookes." After that, he had followed the literature relating to the nitrogen problem with great attention, and he was well-aware of the attempts to oxidise nitrogen using electricity as a source of energy—for example, the test factory of Charles S. Bradley and D. R. Lovejoy at Niagara Falls in the United States. In 1902, possibly inspired by such information, the engineer Eyde and his colleague Hiorth had acquired on behalf of a group of investors the waterfalls at two places in Norway on the river Glomma, Moss and Rjukan. Eyde was convinced that hydroelectricity had a great future, but he certainly had no plans at that time for the exploitation of the power of these falls. He was, along with Swedish collaborators (this was before Norwegian independence from Sweden), gathering assets for a company, and he was subsequently on the lookout for uses to which he might put these potential hundreds of thousands of horsepower. The supply of electricity for the capital, Oslo, was a possibility, but it was not then commercially viable. So it is no surprise that when he met Birkeland in February 1903, he was immediately struck by the idea that he had a source of energy at his disposal that might be used constructively to fix nitrogen in Birkeland's novel electric arc.

Birkeland's ill-fated demonstration of his electric cannon, described above, had foundered because of a violent short-circuit that had produced a spark and a flame. Evidently, the dinitrogen had caught fire! Eyde was convinced that if an energetic electric discharge could be produced in a confined space, then it might be possible to maintain such a flame and trap the nitrogen oxides so produced. The dinner at which Eyde discussed the project with Birkeland was on the evening of February 13, 1903, and only one week later they made a preliminary application for a patent. They also decided to set up a company to exploit this idea.

The magnitude of the problem that Birkeland and Eyde had undertaken to solve was daunting. First, they had to design a furnace hot enough and stable enough to heat air to a temperature of about 2000 °C. They then had to cool the gas issuing from the furnace so rapidly that no significant product decomposition would occur. This reaction of dinitrogen with dioxygen, just like that of dinitrogen with dihydrogen, is reversible, and slow cooling of the gas issuing from the furnace would allow the reverse reaction to occur, regenerating dinitrogen and oxygen. Rapid cooling implies a rapid disposal of heat energy, and this was ultimately to prove the weakness of the whole process. The nitrogen(II) oxide had to be allowed to react with more dioxygen to generate nitrogen(IV) oxide, NO_2, and finally the mixed oxides had to be dissolved in water to generate nitric acid. Nothing like this had ever been done before, let alone on a sufficiently large scale to be commercially successful.

Eyde was not a man to let grass grow under his feet, and already in April 1903, he was ensuring that sufficient waterpower was available to power the process. By July 1903, funds to design and construct an electric furnace were obtained. The company itself was founded on July 5, 1903, but in the meantime Birkeland had already been testing a furnace at the university. This small furnace probably employed a magnetic field to help stabilise the flame, but it could not function for more than a few minutes before it overheated. By May 1903, he was already working in an industrial workshop at Frognerkilen that offered better facilities. This is where the first stable arc (or rather a set of sparks produced at a rate of about 100 per second) was obtained. The applied voltage was about 5000 volts AC. To the eye, it looked like a circular disc-shaped flame with a diameter of about 20 cm. This shape was imposed by a magnetic field, and the furnace was rated at about 3 kW, later increased to 10 kW. The whole development had been very rapid.

By the end of July, the power consumption had been increased to 28 kW, and on August 7, 1903, the first traces of nitric acid were detected when air was passed through the flame and then quenched in water. By September, a flame of 55 cm diameter could be stabilised, and the power consumption was now 45 kW. After that, the furnaces became larger and the periods of continuous operation longer. The demands on the local suppliers of power became greater. The technical problems were considerable (for example, the electrodes often degenerated under the influence of the high temperatures and voltages), but progress was rapid. By the end of the year, a continuous run of 36 hours was achieved. In February 1904, a new copper furnace was run for 130 hours at a maximum power of 120 kW. This was scarcely a year after the initial dinner party at which Birkeland first discussed the project with Eyde. By October 1904, most of the basic problems had been solved and a furnace having a power of 200 kW had been developed.

This achievement should not be underestimated. No one had worked continuously under such extreme conditions of temperature, electrical pressure, and corrosiveness, and every advance had required the most careful investigation. However, it was now necessary to build a furnace that would be able to operate commercially, and that was a further problem. Nevertheless, Eyde, the engineer-entrepreneur, had been as active in procuring funds as Birkeland, the physicist, had been in developing the Birkeland–Eyde furnace. The Société Norvégienne de l'Azote and the Société A/S Notodden Saltpeterfabrik were both founded during 1904. The Wallenbergs, the Swedish banking family, were heavily involved.

By this time, the industrialists could claim to have facilities to produce 900 kg of nitric acid per kW year and to have protected their property with patents in Norway and Germany. Furthermore, they had plentiful hydroelectricity to power the factory. The furnace had developed a power of 500 kW, so that an annual acid production of 450 tonnes was already possible, though only under the most favourable conditions. This wasn't enough to repay the investment and to support further developments. A test factory was built at Notodden, construction starting in spring 1904, and this had the potential electricity supply of 2000 kW. By May 2, 1905, this factory was ready to start up.

During 1905 and 1906, several furnaces were put to use. The individual yields were never as high as had been hoped, perhaps due to incomplete combustion, to decomposition of the nitrous gas after combustion, or to incomplete absorption of the nitrous gases in water. All these matters were rigorously investigated, and slowly the system became more and more efficient and reliable. A further significant development was that the French bank, la Banque de Paris et des Pays Bas, was encouraged by Wallenberg to invest in the process. After intense negotiations, somewhat overshadowed by the dissolution of the union of Norway and Sweden under the Swedish crown, which took place in 1905, the Paris bank sent a mission of independent foreign experts to Notodden to evaluate the system. Just before this mission arrived, the Norwegians managed to stabilise the production at about 530 kg of acid per kW year, which was reckoned to be a commercial yield. The bank mission reported that the process they had examined was the most important that had yet been constructed for the production of nitrogen compounds from atmospheric nitrogen. This report led to the foundation on December 2, 1906, of Société Norvégienne de l'Azote et de Forces Hydro Electriques, with the power to float its shares on the international market. This company later became the modern chemical giant Norsk Hydro. The future seemed assured.

Eyde had pushed the development of the electric arc process so rapidly because he knew that companies such as BASF were also interested in fixing nitrogen and might well provide considerable competition. BASF had even taken out patents to protect its own ideas. In fact, BASF was already working on a similar process by 1906 and had constructed a furnace named after its inventor, Schönherr. Eyde decided to negotiate an exchange of information with BASF, and a delegation headed by Birkeland and Eyde visited the Ludwigshafen factory of BASF in the autumn of 1906. There they found a working furnace that was superior to their own in some ways. It is notable that BASF was already so far advanced in the development of its furnace but that finally, and unlike the Norwegians, they rejected the system completely and concentrated on another process. In the meantime, a collaboration between BASF and the Société Norvégienne de l'Azote et de Forces Hydro Electriques developed.

The visit of the Norwegians to BASF did not stop their plans at Notodden. By December 1907, the factory contained 24 furnaces, each rated at 750 kW. However, all sorts of problems continued to be experienced, with the walls of the furnaces crumbling, the circulation of corrosive gas in the system causing deterioration, and the generators continually breaking down. It was not until the end of 1908 that all the furnaces were functioning reliably and at full power. At the same time, even larger furnaces were being built at Rjukan (figure 5.4 and figure 5.5).

Figure 5.4. The arc process factory at Rjukan (© Norsk Hydro ASA, reproduced with permission).

Figure 5.5. The first practical furnace house (© Norsk Hydro ASA, reproduced with permission).

The final choice for the furnaces to be employed at Rjukan lay between an even larger Norwegian design and a German furnace. Tests to determine which was superior were to be supervised jointly by the two companies in June 1910. Halfway through the operation, the Germans suddenly withdrew. According to Schönherr, the cooling water was not clean enough and caused a muddy deposit in the cooling system. The Norwegians were not convinced that this was really the reason and claimed victory for their technology.

An agreement was made with BASF to build the first phase of the Rjukan factory with 80% of the power to be used by Birkeland–Eyde furnaces and 20% by Schönherr furnaces. In retrospect, this seems rather strange. In any case, BASF already had an acid factory using Schönherr furnaces that had been working at Ludwigshafen since 1897. The Norwegian system seemed to be the most efficient, and future developments at Rjukan were intended to use only the Birkeland–Eyde system. Finally, the German Schönherr furnaces were demounted and sent to Rjukan. Eventually, in 1911, BASF pulled out of the joint company because they were clearly no longer interested in the process.

An even bigger furnace was entering service at Rjukan. This had the enormous power of 4000 kW and was the ultimate design. It is not clear whether the Schönherr furnaces were ever reassembled at Rjukan, and Birkeland–Eyde reigned supreme.

All this finally happened in 1910. The Norwegian process had come a long way from the 3 kW furnace of 1903 to the 4000 kW furnace of 1910. The future seemed assured. BASF had apparently decided to admit defeat and to act as a minor partner in a joint enterprise. The furnaces continued in production at Notodden and elsewhere in Norway until 1934, when they were forced out of business by a commercially much more successful process. A factory was constructed in Spain, but it never functioned. A further factory did come on-stream in the south of France, but it seems never to have produced more than about 10 tonnes of acid per day, whereas a production of 100 tonnes per day had been achieved in Norway. The factory in France, though supported by the French government, ceased production early in the 1920s. By then, World War I had convinced most industrialists that Germany possessed the secret of a much more efficient nitrogen-fixation process. In hindsight, it seems possible that by 1910 BASF already knew that it had this better fixation process at an advanced stage of development and it was no longer prepared to invest much effort in any oxidative process. This brings us back again to Fritz Haber.

The Cyanamide Process

The cyanamide process[19] was the second significant industrial process for fixing nitrogen in our list. It differed from the two others in that it was to produce a large amount of virtually useless by-product, calcium carbonate, $CaCO_3$. The process involved treatment of the calcium cyanamide, like the barium cyanamide described above, with water:

$$CaNCN + 3H_2O \rightarrow CaCO_3 + 2NH_3$$

It also required a very high temperature to promote the initial reactions of lime with carbon and then dinitrogen. The energy consumption was enormous and, like the arc process, the possibility of heat recovery was rather small. It could only operate at all satisfactorily with a good supply of cheap electrical energy. In addition, at the high temperatures involved (1100 °C), the furnace lining was corroded by the reactants. The initial development of this process again came from Germany. The principal persons involved were again Frank and Caro, and the main company was called Degussa.

The initial emphasis was on the preparation of cyanides, such as sodium cyanide from barium cyanide prepared in an electrical furnace, so the initial observation that barium cyanamide was produced alongside the cyanide was perceived as a drawback rather than an advantage. However, in 1900, Frank showed that calcium cyanamide reacts with steam to generate ammonia, and he also demonstrated that calcium cyanamide could be used directly as a nitrogen-releasing fertiliser. Whether the continual application of this material would finally render the soil unsuitable for some crops was apparently not considered.

It was in 1901 that F. E. Polzenius showed that addition of calcium chloride to the calcium carbide reduced the temperature required for the reaction with dinitrogen from 1100 to 700–800 °C. This was a major advance in that much less energy would now be required and the furnace linings would be much more durable. Figure 5.2 shows a typical modern furnace similar to, but probably smaller than, those that were developed at about this period. For several German companies, this was a time of industrial turmoil, and it was not before 1908 that the cyanamide process began to be applied more generally. The high cost of electricity was a determining factor, and, as with the arc process, hydroelectricity seemed the only sensible source.

By 1912, there were at least nine companies fixing nitrogen by the cyanamide process, all using hydroelectric power and at least one using solid fuel, and it is calculated that they were producing more than 150,000 tonnes of calcium cyanamide per year. The process spread to the United States and Canada, where the forerunner of American Cyanamid started production in 1909. The arc process was considered as a possible alternative in North America, but it was never actually used. By 1914, plants were in operation in many countries, and cyanamide was the major worldwide source of synthetic nitrogen. This was presumably because the necessary engineering, unlike in the Haber–Bosch and arc systems, was relatively straightforward. At the start of World War I, the worldwide nitrogen-fixation industry produced 40,000 tonnes, whereas the total world consumption of fixed nitrogen was 750,000 tonnes. Most of this came from Chile in the form of sodium nitrate, or, as in the case of the United Kingdom, as a by-product of coal-gas production. The amount used in industry was probably around 150,000 tonnes. The balance was used in agriculture.

The Haber–Bosch Process

By 1908, Haber, together with Le Rossignol, was again working on the dihydrogen/dinitrogen reaction. Progress was rapid, and by 1909 Haber had constructed a circulating system that was capable of running continuously at an elevated pressure and hence of producing a steady yield of ammonia. This was when he was able to interest BASF in its potential. He had taken steps to protect his invention with a German patent, which was dated October 13, 1908 (figure 5.6).

The barriers to an industrial exploitation were immense. First, there was no ready, large-scale supply of dihydrogen. Second, no one had attempted to process the explosive gas dihydrogen on such a scale before. Third, the gas had to be circulated over a catalyst at a considerable rate and at high temperatures and pressures, and this had not been achieved on such a large scale before. It was not going to be easy to construct an industrial ammonia-synthesis plant, though it should be noted that some of the problems were similar to those afflicting the Norwegian arc process, and BASF had a foot in each camp. Fifth, the catalyst used by Haber was hardly suitable for industrial exploitation, consisting as it did of rare uranium or osmium compounds. Finally, there was the problem of finding a good source of pure dinitrogen since dioxygen seemed to be a poison for the synthesis.

Figure 5.6. The original basic patents for the Haber–Bosch process (© Corporate Archives, BASF AG, Ludwigshafen am Rhein, Germany, reproduced with permission).

Carl Bosch and Alois Mittasch visited Haber's laboratory on July 2, 1909. Bosch saw the system functioning, and from then on he became deeply involved in the research. For his development of high-pressure industrial chemistry, he, together with Friedrich Bergius, was awarded the Nobel Prize for chemistry in 1931, 13 years after Haber's Prize.[20] The ammonia process was referred to in earlier times as the Haber–Bosch Process,[21] and it is sad that the name of Bosch is often no longer associated with it.

Mittasch was given the task of finding a suitable catalyst for the process because Haber's best catalyst was osmium and the total world supply was so small—only a few kilogrammes—that its use commercially was out of the question. As Bosch later remarked, one explosion of the reactor under pressure would have resulted in the loss of the entire world stock of osmium. Mittasch achieved his goal in an amazing spurt of work that resulted in testing 2500 different catalysts in 6500 runs within about one year. In the end, a modified iron catalyst was developed, and this is essentially what is still used today, or at least was used until very recently.

Bosch's job[22] was to turn the whole concept into an industrial process, and he appears to have worked on several problems at the same time. It is obvious that the resources of BASF were sufficient for him to proceed very rapidly, and he did not have the problem of raising funds that seems to have been a major part of Eyde's function in Norway. The required supply of dihydrogen was obtained using water gas as a source. Water gas is a mixture of dihydrogen and carbon oxides, CO_2 and CO. It was made by passing steam over red-hot coal (for this discussion, coal is essentially a form of carbon, chemical symbol C), both of which were in good supply in Germany:

$$C + H_2O \rightarrow CO + H_2$$
$$C + 2H_2O \rightarrow CO_2 + 2H_2$$

Both these reactions absorb heat and take place at elevated temperatures. To maintain the temperatures, air was passed through the coal bed, alternating with the steam flow, the combustion of the coal heating the whole system. The carbon monoxide was removed subsequently by the so-called water–gas shift reaction:

$$CO + H_2O \rightarrow CO_2 + H_2$$

The carbon dioxide was then removed from the gas by passing it through water under pressure, when the dioxide dissolved leaving clean dihydrogen. Ironically, this part of the process is the only one that has changed substantially during more than 90 years of operation of the Haber–Bosch process. Bosch's fundamental contribution to ammonia synthesis is no longer exploited, natural gas being so much cleaner and easier to handle and transform than solid coal. Today, the preferred method of dihydrogen production is by the steam-reforming of natural gas, which consists principally of methane, CH_4:

$$CH_4 + H_2O \rightleftharpoons CO + 3H_2$$

This procedure is then followed by the water–gas shift reaction to remove the CO. However, the major problem was that no one had ever designed and built an industrial plant to operate at 600 °C and at several hundred atmospheres pressure. Bosch started with the system that Haber used in his final experiments in 1909. That was known to work. The problem was not simply one of making a big tube into which one could pack the catalyst. As the catalyst chamber (now called a converter) became bigger, all sorts of problems arose. The gas flowing through had to be distributed evenly throughout the converter, so that the gas did not flow rapidly and directly through channels in the packing without general contact with the catalyst. The temperature had to be maintained at the constant optimum value throughout the converter. Although the walls of the converter had to be thick to withstand the pressure, heat had to be conducted away as quickly as possible in order to maintain the stability of the system. The problem was the familiar one. The apparatus needed to operate at an optimum temperature. The dinitrogen/dihydrogen reaction is exothermic; that is, it gives out heat. However, the equilibrium becomes more favourable at lower temperatures when, however, the slower rate of reaction means that the equilibrium point is reached more slowly. If the temperature is too low, then the gases do not give the optimum yield, which is a compromise between the theoretical maximum and the minimum required for commercial viability.

In another effort to attain efficiency, the heat carried away in the product gases was used to warm the incoming gases from ambient temperature to reaction temperature, so complicated heat exchangers were necessary. Finally, in a large converter containing many tonnes of catalyst, the sheer weight of the catalyst at the top of the converter would crush that at the bottom, so the gas flow would gradually become blocked. The physical strength of the catalyst material was therefore critical. Ideally, a large plant must be able to run for years because it is not easy or commercially viable to dismantle a gigantic piece of equipment at short, frequent intervals.

As the development proceeded, even more unexpected problems arose. For example, the converters were made of steel that contained carbon. The dihydrogen in the reaction gas stream slowly removed the carbon from the steel at the operating temperatures used, causing the converter to crack. This produced the possibility of an explosion under operating conditions. The final solution was to give the converter a pressure-bearing steel jacket over a thin lining of soft steel to protect it from the dihydrogen.

The converter developed was a complicated piece of equipment, essentially a corrosion-resistant reactor and a complex, controllable heat exchanger. Figures 5.7 and 5.8 show converters developed in 1910 and 1911. Modern plants still have to fulfill similar functions. By 1913, the problems cited above and a multitude of further problems had been solved, and a plant was built at Oppau. Figure 5.9 shows a picture of the Oppau plant, but a true appreciation of what was actually involved can be obtained from figures 5.10 and 5.11. Between 1910 and 1915 (the middle of World War I), the volume of the converters used by BASF increased from 0.12 m^3 to about 44 m^3 (figure 5.10). The weight of the converter increased from 0.3 tonnes to 75 tonnes. The output of a large converter was about 20 tonnes of ammonia per day by the end of the war. However, the efficiency as measured by the amount of steel involved to produce a given weight of ammonia increased dramatically up to the end of the war and beyond, by a factor of 4 by 1916 and by a factor of 15 by 1927 (figure 5.11). The converters were becoming much more efficient. However, small they certainly were not, and modern converters are still enormous and complex machines (figure 5.12). Figure 5.13 shows some maintenance workers together with one of their spanners, and figure 5.14 shows a typical converter of that period being installed. By the end of the war, Germany was producing some tens of thousands of tonnes of "synthetic" nitrogen in the form of ammonia.

The Aftermath of World War I

The various combatants in World War I were reasonably well-aware of the productive capacities for strategic materials of the various countries. Fixed nitrogen, as nitrate, was vital for the manufacture of explosives and also as a fertiliser. In Germany, most of this came from gasworks. Some, but not much, was produced at Oppau by the Haber–Bosch process. Rather more

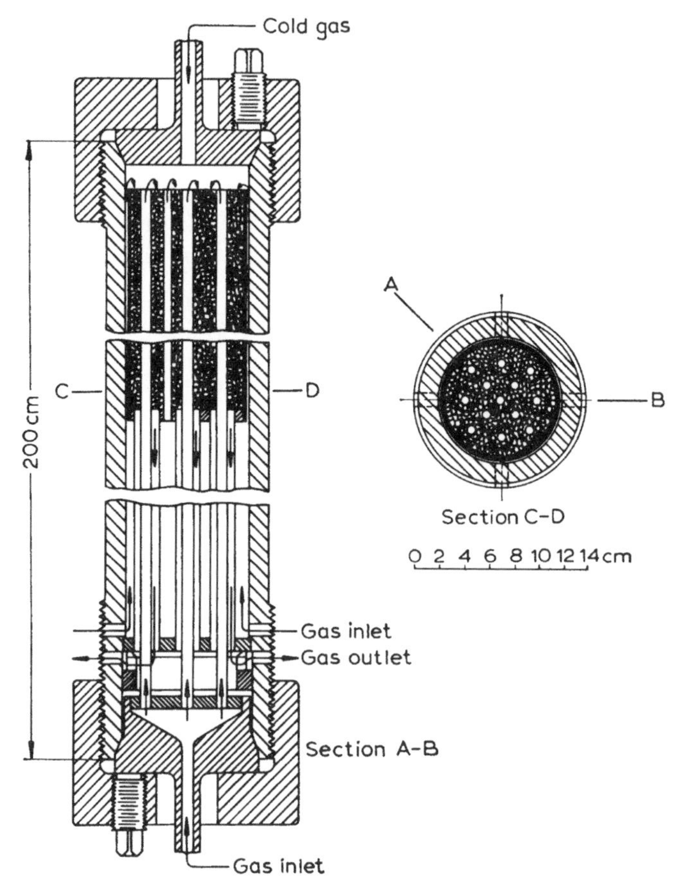

Figure 5.7. A converter of 1910. This converter was about two metres long. It embodied counter-current flow to obtain better temperature control but was ultimately rejected because of the embrittlement problem (© 1931 The Nobel Foundation, reprinted and reproduced with permission).

was fixed by the cyanamide process. The United Kingdom was a very considerable producer of chemicals, though much less so than Germany. Much of the nitrogen used in the United Kingdom was imported from South America, first as guano and later as sodium nitrate. This was perceived to be a sustainable situation, bearing in mind the self-satisfied clichés to the effect that the sun never set on the British Empire and that Britannia, in the form of the Royal Navy, ruled the waves. There was apparently no nascent nitrogen-fixation industry in Britain.

World War I changed this situation. There was suddenly an enormous demand for nitric acid to manufacture explosives, especially picric acid and TNT. Ammonia could be used to manufacture nitric acid by an oxidative process, as shown below:

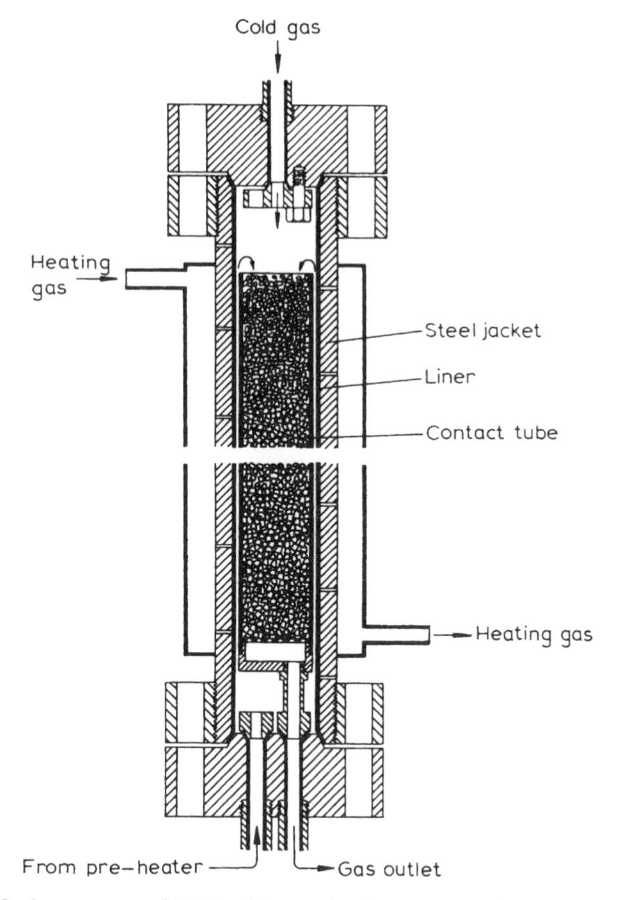

Figure 5.8. A converter of 1911. This contained a protective liner, as described in the text, but the heating was external (© 1931 The Nobel Foundation, reprinted with permission).

$$4NH_3 + 5O_2 \rightarrow 4NO + 6H_2O$$
$$4NO + 3O_2 + 2H_2O \rightarrow 4HNO_3$$

Without a good indigenous supply of ammonia, the only recourse was to the arc process or imported nitrate. The manufacture of nitric acid in Britain during the war reached as much as 20,000 tonnes per year at times. Clearly, this reduced the supply of fixed nitrogen available in Britain for other uses, especially for fertilisers. The situation in Germany was comparable. As L. F. Haber, the son of Fritz by his second wife, stated graphically, "after August 1914 it was not simply a choice between guns or grain, but between guns and defeat."[23]

The problem was particularly severe in Germany, where the import of nitrate from Chile stopped abruptly in August 1914, and this cut the supply of fixed

Figure 5.9. A view of the Oppau plant in 1913. It is evident that no one was very concerned about problems of air pollution, but that was true of the chemical industry everywhere (© Corporate Archives, BASF AG, Ludwigshafen am Rhein, Germany, reproduced with permission).

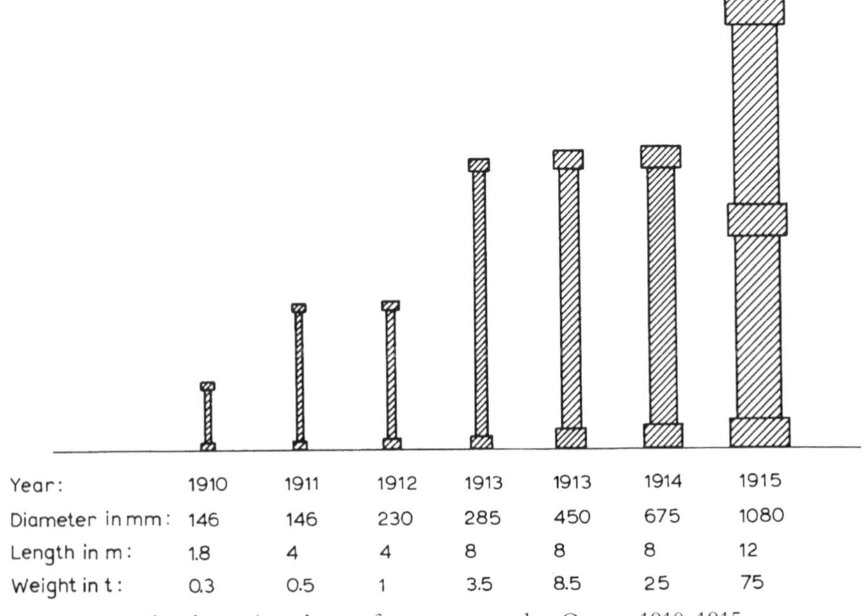

Figure 5.10. The change in volume of converters used at Oppau, 1910–1915 (© 1931 The Nobel Foundation, reprinted with permission).

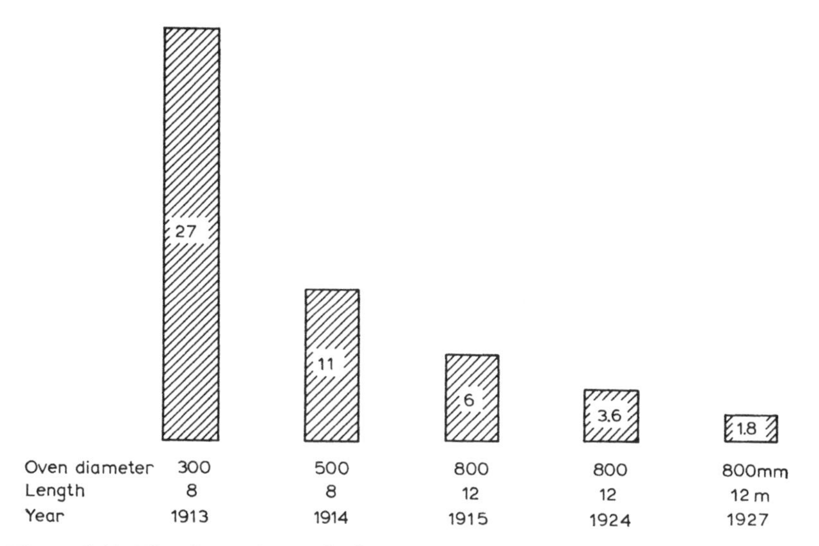

Figure 5.11. The change in weight for a given yield of ammonia of converters used at Oppau, 1913–1927 (© 1931 The Nobel Foundation, reprinted with permission).

nitrogen by half. The result was a severe strain on the German economy. Apparently, von Hindenberg, then a principal commander of the German army, caused an advertisement to appear in German newspapers in 1915. The translation is as follows: "The women of Germany are commanded to save their chamber lye, as it is very needful to the cause of the Fatherland in the manufacture of nitre, one of the ingredients of gunpowder. Wagons, barrels, and tanks will be sent to residences daily to collect and remove the same."

In point of fact, the cyanamide process produced most of the German fixed nitrogen, as ammonia rather than as nitric acid, and considerable efforts went into constructing new plants. The oxidation of ammonia to nitric acid was a well-understood method but was scarcely exploited in 1914. The arc process was under French rather than German control, and Norwegian carbide was controlled by the British. The Haber installations at Oppau were expanded, and new factories were built at Leuna, so that at the end of the war Germany could fix about 95,000 tonnes of nitrogen per year by the Haber process, 50% of its total nitrogen requirements. This was despite the fact that, as Bosch related in his Nobel lecture, the Allies subjected the Oppau factory to aerial bombardment.

Apparently, the British were able to maintain their supplies from Chile even though the German U-boats caused severe problems, but the Germans were not able to obtain nitrogen from this source. The British used a lot of Chilean nitrate to synthesise nitric acid. They even invested in the ammonia oxidation process (see below). In the traditional British way of muddling through, enough nitrogen in the required forms was found. A committee, the British Nitrogen Products Committee, was set up to resolve the problem,

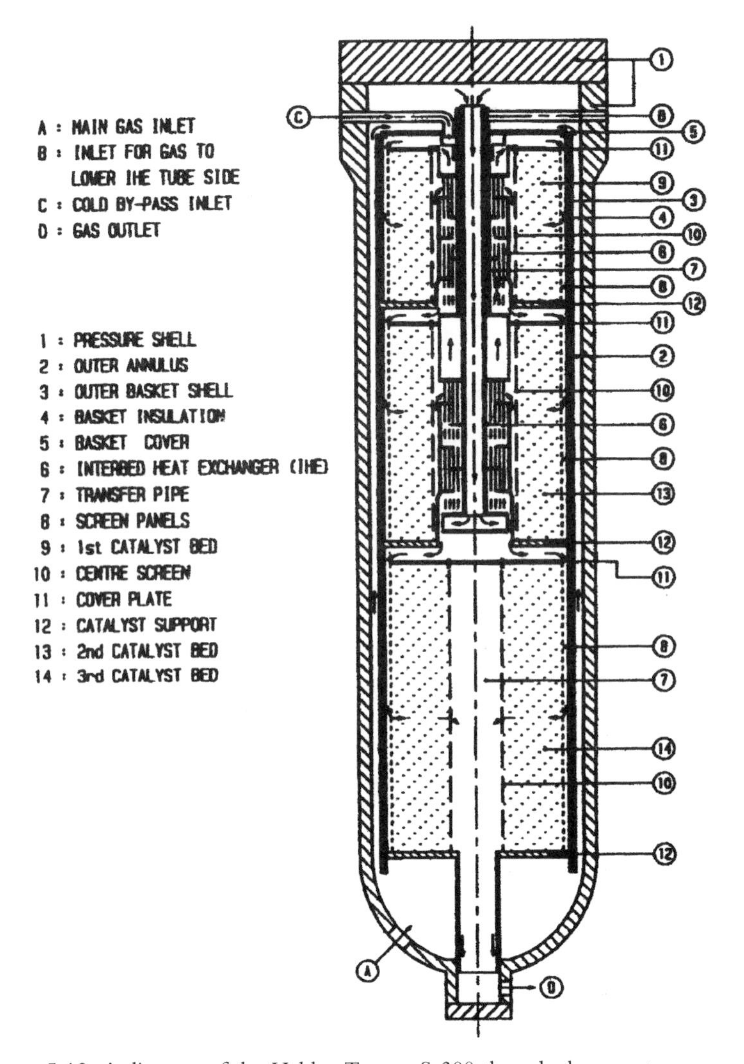

Figure 5.12. A diagram of the Haldor Topsøe S-300 three-bed converter, a modern converter. There are many designs of modern converter, some even designed to be aligned horizontally rather than vertically, but they all perform rather similar functions. A modern plant may produce as much as 2000 tonnes of ammonia per day, whereas Haber in 1909 was producing at the rate of 80 g per hour, or about 2 kg per day (© Haldor Topsøe A/S, reproduced with permission).

and it deliberated during 1917 and 1918. One proposal, to build a cyanamide plant, was never acted upon. Another proposal, to build a Haber-type plant at Billingham in the north of England, had not even started at the end of the war. By then, it was probably not considered worth the effort to repeat the research that Haber, Bosch, and their collaborators had already undertaken.

Figure 5.13. Some maintenance workers together with one of their spanners. It would seem that they are dealing with the top of a converter such as is shown in figure 5.14 (© Corporate Archives, BASF AG, Ludwigshafen am Rhein, Germany, reproduced with permission).

Figure 5.14. A typical converter being installed at Merseburg in 1927/28. The scale is evident from the size of the workmen controlling the installation (© Corporate Archives, BASF AG, Ludwigshafen-am-Rhein, Germany, reproduced with permission).

If the Allies had hoped for Germany to surrender because of a lack of ammunition or food, they were to be severely disappointed. Whatever the strains put upon the German economy, the German army was a formidable opponent for years longer than might have been expected. It was evident by the end of the war that the Germans had a reliable indigenous supply of fixed nitrogen, presumably produced by the Haber–Bosch process, the laboratory details of which Haber and Le Rossignol had published in the scientific press in 1913. However, that was only the beginning, and how industrial production was achieved was not known outside BASF. This may have been an accident of war, which prevented information exchange between the combatants, or because BASF wanted to protect its commercial interests, or both. In any case, the British and the French wanted to know how it was done, and as victors and occupiers of Germany they were well-placed to find out.

The French army occupied Oppau, where the first Haber plant was erected and was apparently undamaged and still functioning. A French variant of the Haber–Bosch process (the Claude process) and an Italian version (the Casale process) were operating by the end of the 1920s, but they required much higher pressures (up to 1000 atmospheres) and temperatures (up to 650 °C) than the original BASF variant of the Haber–Bosch process. Inevitably, they gradually lost out in favour of the German procedure, and eventually the German process was used everywhere. However, the initial problem was to discover how the Germans actually did it.

The British were well-aware of the advantage that Germany had gained from the Haber–Bosch process and also of the near success of the U-boat siege of Britain. A lot of Chilean nitrate was dissolved in the sea while it was in transit to Britain, and it was probably due to the large amount of ammonia by-product from coal-gas works that Britain did not actually run out of fixed nitrogen. Again quoting the economist L. F. Haber,[23] a British government report called the Haber–Bosch process the "key to Germany's war production of explosives," and "for military security it is essential that each country should have its chemical industry firmly established, and this must be secured as one of the conditions of peace, as otherwise we are leaving Germany in possession of a weapon which will be a permanent threat to the peace of the world." Whatever the truth of this kind of statement, it seems to echo through time, right down to the present day, with ideas of mutually assured destruction, weapons of mass destruction, long-range missiles, and nuclear weapons.

The detailed reviews of Creighton[3] concerning industrial nitrogen fixation in 1919, written at the end of World War I, are illuminating. Though he wrote that the nitrogen-fixation problem had been solved and listed many possibilities, he also placed much emphasis on the arc process, and his discussion of the Haber–Bosch process is rather uninstructive. He clearly could not have known much of the developments at BASF during the war, and he says so: "While these difficulties [concerning temperature, pressure, etc.] have been overcome, the details have been kept secret. Also there is but little information regarding the details of the construction of the furnace and its operation." Creighton notes that world synthetic ammonia production by the Haber–Bosch process was cheap ($0.04 per pound) and reached 140,000 tons in 1918. He also states that synthetic ammonia was being produced, as well as in Germany, in the United States and in "England." In the last instance, he was probably wrong.

What followed in Germany at the end of World War I was a kind of foretaste of what happened at the end of World War II when the Soviets and the Western Allies raced into Germany as far and as fast as they could, partly in order to gain control of as many German scientists and as much of their knowledge as possible. The British at least wanted badly to know in detail how BASF fixed nitrogen. The philosophy was enunciated quite openly by Major F. A. Freeth, who became an important figure in the postwar development of ICI.[24] He is reported as saying on November 12, 1918, the day after the Armistice was signed, to Lord Moulton, who was an eminent judge, as well as being chief of the Department of Explosives Supply, and who was commendably discomforted by the statement of Freeth, "Sir. In three weeks' time the British army will be on the Rhine. Let's send a chemical commission and pinch everything they've got." This could have meant that the British would be able to adopt the German technology without paying for it, and whether the Germans liked it or not. This is not quite how it actually happened.

The British effort on the Haber–Bosch process did indeed get going during World War I, but no full-scale plant was constructed before the end of the war. As we have seen, in Germany it had taken the genius of Bosch to achieve that. Although land was set aside by the British Government at Billingham, by 1919 it was persuaded that this whole undertaking was better managed by private industry and tried to wash its hands of the project. The company that eventually stepped into the breach was Brunner, Mond & Co., the forerunner of ICI. Even they were not prepared to finance the whole development from their own resources. In any case, in the absence of the demands of war, there was enough fertiliser for agricultural use available from gasworks. Under pressure, Brunner, Mond & Co. agreed to go ahead at Billingham, but part of the deal seemed to be that they could take a look at the installation at Oppau, something that had already been done rather superficially at the end of the war.

The official history of ICI tells what happened in rather a dry fashion.[24] Oppau was in the French zone of occupation. BASF was determined to expose as few of its secrets as possible. The company told the French authorities that if the British were allowed to inspect the factory, then the works would be closed and the 10,000 employees would be laid off. Feeding 10,000 unemployed men and their dependants was a problem the French did not want to confront. In any case, there was a suspicion that the French were quite happy to watch the factory operation from afar and to glean what they could to enhance their own factories.

Eventually, a British mission from Brunner, Mond & Co. was allowed to go anywhere and see anything at Oppau but not to measure or make pictures.

The visit took place in May and June 1919 and was a disaster from the British point of view. The delegation was boarded in the French officers' mess. For 25 days, they "lived on bread and wine alone with not even a pat of butter." The French apparently did not think to provide any roast beef. The works were fully manned but not functioning, most of the staff being happily engaged in repairs and maintenance. Apparently, the BASF management were no keener for the French to see the plant start up than they were the British. The official ICI history states the situation very clearly: "All concerned did their best to frustrate the British. They took ladders away to stop them climbing on to raised platforms. They took away gauges or blacked out the glass . . . If they went into a building where there was work going on, it immediately stopped and everyone stared at the intruders until they went away again. The members of the mission [were] forced to rely entirely on observation and memory."

Even worse was to follow. At the end of the stay, all the baggage, including most of the written report, was placed in a locked railway wagon under control of an armed guard, who was reported to be an elderly Frenchman. Somehow, someone cut his way undetected into the wagon through the floor and removed the contents. This was discovered only when the wagon arrived in Paris, en route to Britain. It is not stated what became of the armed guard. The report had to be rewritten after the return to Britain using memory and some documents that had not been stored in the wagon.

There were interminable delays in constructing the new plant at Billingham, at least in part because BASF was still not keen to vouchsafe precise details, even in commercial agreements. It had a virtual monopoly on the Haber–Bosch process, which it was determined to maintain. Its production in 1920 was about 1.5 million tonnes of fixed nitrogen, whereas by 1926 only about one-thousandth part of this was fixed at Billingham. The production there reached 100 tonnes per day only by 1928.

The real breakthrough for the Billingham works was provided by the appearance of two engineers who claimed to have been trained at Oppau and to have supervised the installation and start-up at Merseburg (Leuna).[24, 25] These two men, Koebele and Adler, offered all the details of the plant to Brunner, Mond & Co., but for a price. They had apparently already sold similar information to a French company, and after some hesitation the British also paid for it. Apparently, Koebele and Adler were genuine and their information was good. The British had all the information they needed, and progress was much faster. Whether Brunner, Mond & Co. had behaved entirely ethically is a matter of opinion. The whole matter bears a striking similarity to the way in which the secret of porcelain manufacture finally leaked out from Meissen to Sèvres about 200 years earlier. Industrial espionage is no new phenomenon.

The adoption of the Haber–Bosch process by the British effectively spelled the end of the Chilean nitrate industry, but this was not the only consequence. In his introduction to Haber at the Nobel Prize presentation in 1921, the President of the Royal Swedish Academy of Sciences compared the relative costs of the production of calcium nitrate by the various processes then extant, the direct arc process, the Haber–Bosch process followed by oxidation of the ammonia, and the cyanamide process followed by hydrolysis and then ammonia oxidation. These were in the ratios 1.00:1.03:1.17. In addition, the Haber–Bosch process did not rely on cheap hydroelectricity and did not produce unwanted by-products other than some carbon dioxide. It was already clear by then that the Haber–Bosch process was the best unless the particular products of the cyanamide process were required.

In fact, the cyanamide process lived on healthily for quite some time. In 1910, world production of cyanamide amounted to 20,000 tonnes. World War I encouraged expansion, so this rose to 600,000 tonnes. The process was gradually improved. During World War II, Germany alone produced 3 million tons, used for agriculture and for plastic and acetylene production. The end of that war spelt the end of much of this, though in 1947 there were 39 cyanamide plants in 17 countries with a capacity for 1.5 million tonnes of cyanamide. Production was only about half of this capacity. However, the life of the arc process was much shorter.

The most important factor in favour of the Haber–Bosch process was that it used about 25% of the energy of the arc process. The latter could only be commercially viable with a cheap source of electricity. Attempts were made until 1925 to adapt the Schönherr process to run at slightly elevated pressures, which certainly increased the yield of acid. It was also realised that the arc process would never be able to compete with the Haber–Bosch process given an equal cost for energy. Norsk Hydro was obliged to change horses in midstream.

It was widely understood by the end of World War I that one might use high operating pressures, as did the Claude and Casale processes, or pressures as low as 90 atmospheres, as in the United States, or anything in between. However, the catalysts were generally closely guarded secrets. BASF was the only company with functioning factories, and these operated under conditions somewhere between the two extremes. The Norwegians did not have the advantage of the Allies of being able to inspect the BASF plants in detail. They also had no significant coal deposits. However, they still had cheap electricity.

Norsk Hydro had to reinvent the Haber–Bosch process virtually for themselves. Like Bosch, they developed methods for preparing and purifying dihydrogen from coal, but they also used water electrolysis. This was the source employed in their new plants. They investigated new catalysts and finally adopted iron, probably much like the BASF catalyst. The one they used was produced by collaboration with the Nitrogen Engineering Company of the United States. In 1923, the Norwegian government was pressing for an expansion of the nitrogen industry, despite very difficult economic circumstances. There was considerable disagreement as to how to proceed. Finally, in 1927, an agreement was reached with the successor of BASF, IG Farbenindustrie, to convert successively the arc installations at Rjukan to Haber plants. The

German company would allow the use of its patents, due to expire shortly, and would also render technical support. Tests of a small installation of the Nitrogen Engineering Corporation were successfully completed in 1928, and these tests were the basis of the ultimate installations. In 1929, the arc process virtually ceased production. Even so, in a patent filed in 1925, Norwegian industrialists described a process for producing ammonia using an electric arc via an intermediate, hydrogen cyanide, HCN. The arc technology did not die without a fight!

The Manufacture of Nitric Acid

Like ammonia, nitric acid has a long history. The acid and its salts, the nitrates, were for centuries recognised as compounds of value, the acid because of its unique property when mixed with hydrochloric acid of being able to dissolve the most noble of metals, gold, and the salts because of their use in explosives and later in fertilisers. The Arabian chemist Geber described the preparation of nitric acid from saltpetre in the twelfth century, and alchemists and chemists experimented with it throughout the succeeding centuries. However, until its constitution was finally established in the eighteenth century, no alternative synthesis could be easily achieved. Berzelius stated that ammonia could be burned, and the optimum conditions for catalysis and to produce specific products were established by the efforts of a large number of workers. Amongst the more successful attempts to commercialise the oxidation was that of Frank and Caro, whose work on the cyanamide process has already been described. They used a furnace and catalyst that are essentially like those that are used today. In summary, the fact that ammonia gas can burn in air to yield oxides of nitrogen and steam has been known for about 200 years. These can then be absorbed in water to yield the required acid. The sum of several reactions may be represented by the following equation.

$$NH_3 + 2O_2 \rightarrow H_2O + HNO_3$$

This overall process gives out heat, and it also results in most of the dihydrogen so expensively prepared for the synthesis of ammonia being converted to water. This suggests that it might be helpful to integrate this process with the Haber–Bosch process in order to be able to reuse some of this energy but also that if someone could come up with a process to make nitric acid that didn't involve the formation of unwanted water, then the preparation of dihydrogen would not be necessary and the production of 100% nitric acid at one stroke might become a viable industrial option. This still has not been achieved.

At the beginning of the twentieth century, the nitrate imported from Chile was an excellent source of nitric acid, at least for Britain and the United States, and for others in time of peace. The arc process gave nitric acid as its principal product and seemed destined to provide an alternative supply of nitrates, but it could not compete commercially with the Haber–Bosch process, at least for primary production of fixed nitrogen. In the absence of an adequate supply of nitrate from Chile, and due to the commercial failure of the arc process, an alternative source of nitrate had to be found. Since ammonia finally became the only major source of fixed nitrogen, it was necessary to develop a synthesis of nitrate from ammonia. Fortunately, this was not difficult to find.

Mellor,[4] writing in the 1920s, listed three manufacturing processes for nitric acid. One is the arc process, as discussed above, but this did not survive beyond the 1920s. The second was the sulfuric acid/nitrate reaction, which had been worked up from a laboratory process. This relied on the supply of Chilean nitrate, and that was not assured. Indeed, writing in 1923, Cottrell[26] described in detail the industrial manufacture of nitric acid from caliche, the naturally occurring form of sodium nitrate, and noted that the extraction process was highly inefficient and that reserves, then estimated to be 240 million tons, were substantial though not inexhaustible. The exports of Chilean nitrate peaked at about 3 million tonnes during World War I but were already less than 1 million tonnes by 1920.

The final method for nitric acid production was the oxidation of ammonia. This is the method that has survived. Modern plants for the oxidation of ammonia are rather complicated systems. The reaction is carried out in a reactor containing a catalyst consisting of layers of fine gauze of platinum or platinum/rhodium. However, because of the possibility of side reactions giving unwanted products, the temperatures must be carefully controlled. Figure 5.15 shows the layout of a typical plant, and figure 5.16 is a picture of the gauze in the catalyst chamber. This will normally be about 2 m in diameter and hold up to 30 layers of catalyst gauze, maintained at about 900 °C. Nitric acid manufacture is not a poor person's option! It is also notable that the dihydrogen so expensively produced for the manufacture of ammonia is here ultimately converted to water. This may stimulate ambitious researchers to find ways to produce nitric acid not involving oxidation of ammonia!

The manufacture of nitric acid takes place on an enormous scale. Typical plants are shown in figure 5.17 and 5.18. In the United States in 2000, about 9 million tonnes of 100% nitric acid were produced. Much of this acid was converted to ammonium nitrate by reaction with ammonia from the Haber–Bosch process, which is another reason for the integration of ammonia and nitric acid manufacture:

$$HNO_3 + NH_3 \rightarrow NH_4NO_3$$

Even this simple reaction must be achieved in an enormous and complicated plant (figure 5.19). A modern ammonium nitrate plant is shown in figure 5.20. Ammonium nitrate is much used as a fertiliser and also as an explosive. In the year 2000, the production of ammonium nitrate in the United States alone was about 8 million tons.

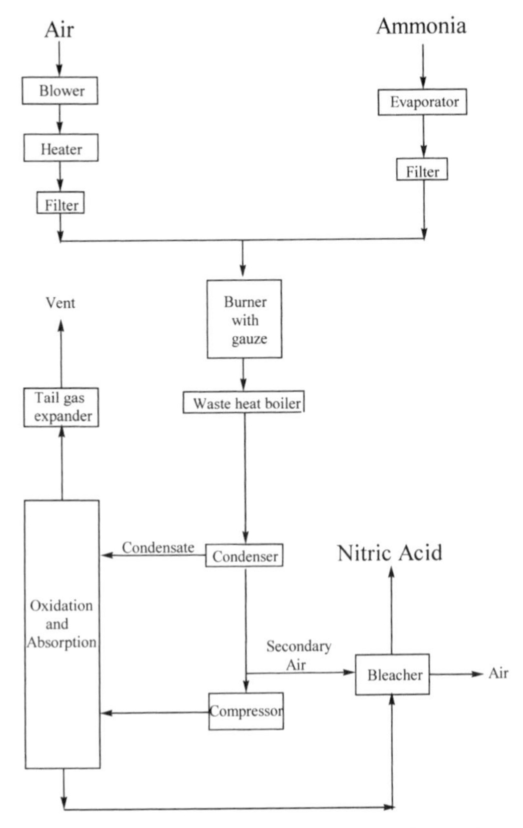

Figure 5.15. Flow diagram of a dual-pressure nitric acid plant. This is one of several possible systems. The air and ammonia are cleaned and combined and burnt in the catalyst chamber. This produces a mixture of oxides of nitrogen, dioxygen, and water, and some heat, part of which is recovered. The condensate passes down the absorption tower, where most of the nitric acid is formed. The non-condensable gases from the burner are passed up the absorption tower, but part is used to purge NO_2 (nitrogen dioxide) from the acid using the bleacher. Nitrogen dioxide in the air forms the brown plume often seen over nitric acid plants. For more details, see M. V. Twigg, *Catalyst Handbook*, 2nd edition, Wolfe Publishing Company, London, 1989.

The Nitrogen Problem Solved?

The real benefit from all this work and all this investment was that, theoretically at least, the problem of supplying enough fixed nitrogen to feed the world's population had been solved, and only twenty years after Crookes had brought it to the attention of the world at large. It can be seen as the culmination of millennia of work and study by peoples and civilisations all over the world. There was potential to feed a population much larger than that which then existed. However, the solution really was one adapted only

Figure 5.16. The interior of an ammonia-oxidation catalyst chamber, the ammonia burner. The catalyst chamber is about two metres in diameter. The catalyst itself consists of layers of fine net made from precious metals such as platinum and rhodium. Surprisingly, this catalyst is actually slightly volatile at the high temperature produced by burning the ammonia and is slowly lost in the gas stream. Reproduced by permission of Johnson Matthey Plc.

to those wheat-eating peoples who lived in a successful capitalist and commercial economic system. The racial terms in which Crookes had expressed his warnings seemed to have determined the form in which the solution was found, perhaps not surprisingly. Only an advanced and technically adept society could have framed it. Nevertheless, in the long run, the "solution" was going to bring problems of a different kind that could not then have been envisaged. These are discussed in chapter 7.

Since the 1930s, the Haber–Bosch process has been the only significant industrial method of fixing nitrogen. In all its essentials, it is basically the same process as that developed by Haber and Bosch. The only difference is that dihydrogen is now produced by steam-reforming of natural gas rather than from coal or by water electrolysis, even in Norway. The catalysts have been improved. Rotary compressors have replaced the enormous reciprocating compressors of the 1920s. The size of the plants has continued to grow, so that a modern installation will produce up to 2000 tonnes of ammonia per day and will cost hundreds of millions of dollars. As discussed above, a typical plant layout is sketched in figure 5.21, and a modern plant is shown in figure 5.22, which is a variant on the picture shown in figure 1.8. The efficiency of such plants is close to the theoretical maximum, and ammonia is cheap, at least for Western farmers. The worldwide fixation of nitrogen

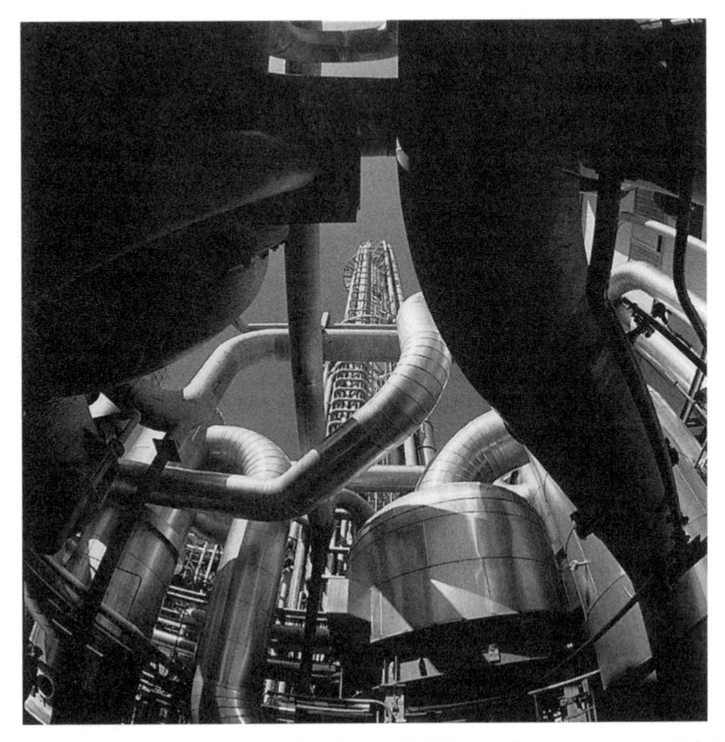

Figure 5.17. Part of a nitric acid plant in the BASF complex at Antwerp, Belgium. This picture shows clearly the ammonia burner and an absorption tower. Nitric acid production is often integrated with ammonia production, which occurs in a Haber plant like that shown in figure 5.22, to allow the immediate synthesis of ammonium nitrate in a plant like that shown in figure 5.20 (photo: BASF, reproduced with permission).

industrially is of the order of 100,000,000 tonnes per year, not very different from the estimated magnitude of biological fixation. The world food problem is solved for those who can afford to buy ammonia. It still remains for those who cannot.

It is now apparent that any price for ammonia above zero is too high for subsistence farmers in much of the world. Some other way has to be found to improve their crop yields and soil fertility. An imaginative application of the arc process was developed in the 1970s by the Kettering Foundation laboratory at Yellow Springs, Ohio.[27] The basic apparatus was two sewer pipes, one of which was fitted with a pair of iron electrodes. Power was supplied by solar panels, waterpower, or any other suitable renewable source. A cheap truck alternator provided a spark in air between the electrodes. This generated oxides of nitrogen in the way known to science for at least 200 years. The oxides were then driven by a fan through the second pipe, which was filled with lime. The nitrogen oxides formed nitrates and nitrites with the lime, and after a suitable time, the pipe could be broken

Figure 5.18. Another view of a nitric acid plant, this time taken from outside the complex. The size can be judged from the vehicles visible at top left. The towers are absorption towers, and the ammonia burner and associated equipment are inside the sheds. Reproduced with permission (copyright © Kemira Growhow UK Ltd.).

open and the nitrated lime used as a fertiliser. The whole setup could be manufactured for a few hundred dollars, would not consume expensive electricity, would be very robust, if inefficient, and could even be used to heat a hot plate for cooking.

Although it was field-tested, the Kettering apparatus has never received general manufacture and dispersion. This merely emphasises that solving world food problems is not technically difficult. The technology is available, but we do not know how to apply it widely. World hunger is a social, economic, and political problem, and science alone will not provide the complete solution. In fact, it may do more harm than good.

The Fate of Haber and Bosch

Carl Bosch continued to work for BASF and for its successor, I. G. Farbenindustrie. Together with Friedrich Bergius, he developed a series of chemical processes that involved high-pressure syntheses, including those of methanol and the hydrogenated oils. Today, high-pressure chemistry is a standard function of the chemical industry. Bosch became a director of BASF

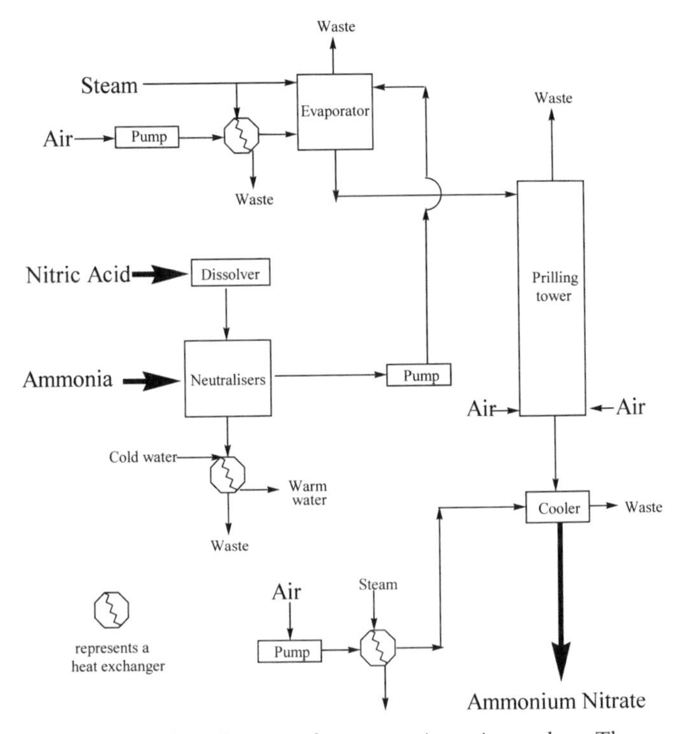

Figure 5.19. A simple flow diagram of an ammonium nitrate plant. The neutralisation of nitric acid with ammonia produces a great deal of heat. This is recovered in a series of heat exchangers. The heat is enough to evaporate off all the water in the ammonia and nitric acid and that produced in the neutralisation. The final drying is carried out by releasing a spray of concentrated ammonium nitrate solution at the top of the prilling tower against a current of air. The ammonium nitrate is solid by the time it reaches the bottom of the tower. The temperature is carefully controlled at all times, in part to recover as much heat as possible and in part to avoid overheating the solid ammonium nitrate, which might explode. For more details, see R. Thompson, editor, *The Modern Inorganic Chemicals Industry*, Royal Society of Chemistry, London, 1977.

and received many academic honours, both in Germany and abroad. In 1937 (after the Nazis came to power), he became President of the Kaiser Wilhelm Society, an organisation that, like the Max Planck Society today, ran several large research institutes throughout Germany. He died in April 1940, at the age of 65.

The life of Fritz Haber seems to have been much less happy. He clearly had a phenomenal capacity for work, was reckoned to be a charming man with a great capacity to see to the bottom of complex problems, and had a forceful personality. He was of Jewish descent and was also an ardent German patriot. During World War I, after he had essentially set the basis for

Figure 5.20. A plant for the preparation of ammonium nitrate. The very high prilling tower and evaporator is clearly visible (75 metres tall). Behind it is the ammonia storage tank, which can hold 10,000 tonnes. In the front centre is a plant for making NPK (nitrogen/phosphorus/potassium) fertiliser, and to the right of the prilling tower is a nitric acid facility. Reproduced with permission (© Kemira Growhow UK Ltd.).

industrial nitrogen fixation, he promoted in Germany the idea that chemistry could not only sustain the German war effort but could help bring the war to a rapid conclusion.

Haber's idea was to use poison gases such as phosgene and chlorine to attack the enemy. The effects would be so terrible that everyone would appreciate the senselessness of the carnage and quickly make peace. Some of this, at least the horrible nature of the weapons, was true. However, the British were not slow to retaliate in kind. The weapons themselves, including the mustard gases that were then developed, were unreliable and sometimes actually more damaging to those who deployed them than to the enemy. Their use and effectiveness were critically dependent on the weather and climatic conditions. The terror they caused was so well-remembered that these weapons were not deployed in World War II in Europe, though gases have been used elsewhere, both before that war and since.

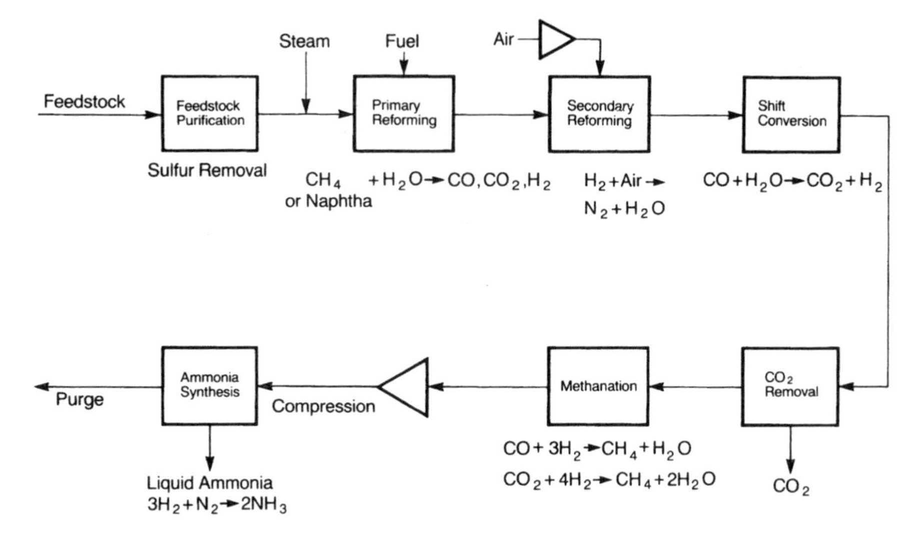

Figure 5.21. A typical modern Haber plant layout using steam-reforming to produce dihydrogen. This is one of several possible arrangements, and the methane (CH_4) or naphtha are two of several feasible feedstocks. The feedstock is first purified (sulfur removal), then reformed in two steps to give a mixture of dihydrogen and carbon oxides (which must be removed); then the carbon monoxide is converted to the dioxide (shift conversion), the carbon dioxide is removed by dissolving it in water or alkali under pressure, the last traces of monoxide are removed by methanation, and finally the gaseous mixture (now three parts dihydrogen to one part dinitrogen) is converted to ammonia. It will be evident that a large part of the cost of this plant is for equipment to produce clean dihydrogen. Reproduced with permission from J. R. Jennings, *Catalytic Ammonia Synthesis*, Plenum, New York, 1991.

For Haber and his family, the outcome was tragic. His first wife, Clara, who also had a doctorate in chemistry, was very opposed to the idea of using chemistry in this way. Whatever Haber's ultimate justification for pursuing his ideas, his wife could not accept it, and she committed suicide. Even after World War I, the award of the 1918 Nobel Prize for chemistry to a man considered in some circles to be a war criminal caused considerable argument. For a variety of reasons, the Prize itself was not actually bestowed until 1920.

After World War I, Haber was made director of the Kaiser Wilhelm Institute for Physical Chemistry and Electrochemistry in Berlin. As a German patriot, he continued to support the state to the best of his ability. One of his projects was to recover gold from seawater to help pay for German reparations to the Allies. It failed. However, he made his Institute one of international renown that attracted many famous foreign scientists. After 1933, when the Nazis came to power in Germany, they felt they did not

Figure 5.22. A modern Haber plant. This is the ammonia plant of BASF at Antwerp, Belgium. The product is principally used for fertilisers. The complexity of columns, furnaces, and pipelines is clearly evident. Such plants are enormous, as can be judged from the figures in the foreground (photo: BASF, Ludwigshafen-am-Rhein reproduced with permission).

need the services of a man who had contributed so much to Germany but who was, unfortunately, originally a Jew. He resigned his offices after being ordered to dismiss any racially undesirable staff members and died in Basel, Switzerland, in 1934. His ultimate goal was asylum in England. The bitterly disappointed and humiliated progenitor of the basis for modern intensive agriculture was 65.

CHAPTER 6

The Continuing Mystery of Biological Nitrogen Fixation

Fixation and Nitrogen Fertilisers from 1920 to the 1960s

Fritz Haber was not blinded by his own achievements. Even in 1921, just when his own work was becoming widely recognised, he could state: "It may be that this solution is not the final one. Nitrogen bacteria teach us that Nature, with her sophisticated forms of the chemistry of living matter, still understands and utilises methods which we do not as yet know how to imitate."[1] The realisation that this was the case prompted a lot of speculation, but scientific advances in biological fixation still awaited a strategic breakthrough. In addition, chemistry, and especially inorganic chemistry, went into decline. The academic world seemed to believe that the chemistry of simple species such as dinitrogen was completed. There was no single clarion call, comparable to that of Crookes in 1898, for the regeneration of research into nitrogen fixation, but the pressure for it built up in a variety of unexpected ways. Perhaps the seminal influence on the field arose in the 1960s.

The stimulus can be seen in the changes that occurred after World War II. In 1945, when much of the world was on its knees, having sustained grave losses of material and people, the impetus was to restart and rebuild. By about 1960, there was the appreciation in some areas that technology could not be applied to the environment indefinitely, nor could standards of living continue to rise without some unpleasant consequences. Western governments were not keen to hear such ideas. The British government, for example, produced a policy document called "Food from our own Resources," the aim of which was to guide the United Kingdom towards self-sufficiency in food, avoiding the possibility of the country being starved as a result of a siege of the sea lanes by a potential enemy. The difficulties of maintaining food supplies from the Empire in the face of a sea blockade were a principal reason why food had been severely rationed in Britain during World War II. A consequence of this policy was that farmers were encouraged to produce as much food as they could, by whatever methods seemed most appropriate, and the era of intensive agriculture really got into its stride. This certainly involved the widespread application of nitrogen fertilisers (and, incidentally, of pesticides, herbicides, and fungicides).

However, warnings that this might not be altogether sensible were also beginning to appear. Like many other reform movements that arose during the 1960s, the ecological movement was becoming active. A major influence was the Club of Rome, a group of scientists, economists, educators, industrialists, and civil servants that first met together in 1968. They were convinced that the problems then facing mankind needed to be confronted in a global fashion. They commissioned a report on what they saw as a major predicament posed by poverty, the degradation of the environment, alienation, growth of cities, and so on. They invited a group at the Massachusetts Institute of Technology (MIT) to analyse these issues using a global computer model, concentrating on five particular issues: population, agricultural production, natural resources, industrial production, and pollution. Their report was published in 1972 and was called *The Limits to Growth*.[2]

The conclusions described in that report were contested, and the global model was severely criticised, but the general tenor of their conclusions, that there are indeed limits to economic growth, cannot really be denied, though some today still seem unable to accept it. The major criticism of *The Limits to Growth* seems to have been that simple extrapolation of exponential growth using an approximate computer model of the world that could not have been accurate was bound to lead to false inferences, particularly as to timescale. *The Limits to Growth* showed how the consumption of fertilisers of all kinds was increasing exponentially in 1970, as were the urban populations in underdeveloped countries and the world population as a whole. Increases in the demands for raw materials, including metal ores and sources of energy, and the growth in pollution by species such as carbon dioxide and lead salts arising from industrial exploitation, were clearly not sustainable. The report also suggested how an equilibrium might be achieved, but this was not presented as a kind of futurology. It was more a warning that something needed to be done and on a global scale.

Thus, about fifty years ago, two trends in thinking about the environment and the related economic pressures were evident. On the one hand, governments were attempting to mobilise all their resources to promote economic growth to meet the newly revived expectations of their populations. They were not inclined then (as some governments even today are not inclined) to listen to warnings that there were limits to their possibilities. This was not a message to appeal to their electorates. Increased food production was often a priority. In addition, there was a realisation that the people of the developing world could not be excluded indefinitely from the benefits enjoyed by the developed world. On the other hand, there was a growing feeling that exploitation of Earth for short-term ends without consideration of the long-term results was bound to be disastrous sooner or later. The results of the conflict of these two pressures will be described in the next two chapters.

Biological Nitrogen Fixation Unravelled (Almost)

The Enzyme that Fixes Nitrogen

By the beginning of the 1960s, there was a widespread feeling that technical innovation and basic research were tools that would solve most problems facing humanity. There was a lot of money available for investment in research, and the nitrogen problem in agriculture was a very prominent feature that received a lot of emphasis. At a scientific level, the kinds of question being asked were: Why is it that legumes are preponderant amongst the plants that participate in symbiotic nitrogen fixation? Can we produce other nitrogen-fixing plants, especially cereals? Can we make cheaper nitrogenous fertilisers? How does biological nitrogen fixation work at the atomic level? Can we adapt the chemistry of the biological system to other ends? Some of these questions were rather misplaced, as was the furore in the popular press that sometimes accompanied relatively minor advances in the science. For example, whether fertiliser is cheap or expensive is a subjective question. The answer is determined, to a large degree, by the economic condition of the buyer, the farmer wishing to use it. By most Western standards, nitrogen fertiliser was indeed cheap already. Even today (2003), ammonium nitrate fertiliser costs less than about $90 per tonne, and the market is enormous, of the order of $1,000 million per year in the United States alone. In any case, since very few fundamental advances in this kind of chemical and biological nitrogen science had been made since about 1930, something dramatic was needed to reopen the scientific and industrial quest for something better.

In 1960, the necessary dramatic breakthrough in the study of biological nitrogen fixation was reported. The first reliable and reproducible method for extracting the enzyme nitrogenase, the enzyme responsible for mediating biological nitrogen fixation, was announced by Carnahan, Mortensen, Mower, and Castle from the chemical company Dupont.[3] Until this was achieved, the study of biological nitrogen fixation tended to be restricted to the study of whole organisms. What is going on inside the cell cannot be deduced with any degree of precision from such observations. For example, one might change the air pressure over a cell and see some effect upon the rate of ammonia formation. However, it is rarely possible to deduce from observations on whole cells how this effect might operate at a molecular level.

The cell-free extracts of nitrogenase were obtained from dried preparations of the anaerobic bacterium *Clostridium pasteurianum*. The availability of cell-free extracts opened the era of detailed studies. It was quickly established that the enzyme from *C. pasteurianum*, not surprisingly, is very air-sensitive, so all the studies had to be carried out under strictly anaerobic conditions. This was later found to be the case for all nitrogenases, whether derived from symbiotic rhizobia or from free-living bacteria and whether they are anaerobic or aerobic. Cell-free extracts of several other nitrogenases were obtained shortly afterward.

Subsequent studies showed that the enzyme is not a single protein. All the known nitrogenases consist of two (or occasionally three) major proteins, and these are generally separable by established biochemical techniques. One protein has a molecular weight of about a quarter of a million, and analysis showed its major inorganic constituents to be iron, molybdenum, and sulfide ions. Since molybdenum had been shown many years before, in the 1930s, to be necessary for nitrogen fixation,[4] and indeed the degree of nitrogen-fixation activity seemed to be related directly to the molybdenum content, it soon became accepted dogma that this protein carried the active site of the enzyme system and that a molybdenum atom was essentially the atomic site at which dinitrogen from the air was first coordinated (bound) and then reduced to ammonia. This protein is sometimes referred to as the molybdenum–iron protein. The other major protein was found to have a molecular weight of about 60,000, and it seemed (and still seems) to act as an electron carrier, picking up electrons from some source within the cell and transferring them to the larger protein. This protein is sometimes called the iron protein.

Biochemists often add the suffix –ase to a name to indicate an enzyme that operates in a particular way or on a particular material. For example, an enzyme that operates upon hydrogen is termed a hydrogenase. Using this approach, a convenient way to name these two proteins that is used widely by biochemists is as dinitrogenase for the larger and dinitrogenase reductase for the smaller. These names indicate that dinitrogenase transforms dinitrogen, whereas the other protein reduces the dinitrogenase. The combination of the two proteins as they occur in a nitrogen-fixing cell is still termed nitrogenase, the enzyme system that transforms nitrogen, but how this happens at the atomic level was still not evident in the 1960s. The analytical and related data from *Klebsiella pneumoniae* shown in table 6.1 were found to be similar to those from other molybdenum-containing nitrogenases and are representative. Even data from the much rarer variants that seem to use vanadium and iron or just iron in their fixation systems (see below and table 6.1) are not very different apart from the specific metal contents.

How Does Nitrogenase Work?

Nitrogenase can actually reduce a whole range of other substances, collectively termed nitrogenase substrates, and perhaps the most interesting is acetylene, C_2H_2, now more correctly called ethyne. The product is C_2H_4, ethylene (or ethene), and there is usually no sign of further reduction to C_2H_6,

Table 6.1
The general compositions of selected dinitrogenase proteins. There is a striking similarity among all the variants, whatever their source and whatever the precise type: the molybdenum, vanadium, or iron-only varieties.

	Klebsiella pneumoniae dinitrogenase	*Azotobacter vinelandii* dinitrogenase (iron-only variant)	*Azotobacter chroococcum* dinitrogenase (vanadium variant)	*Klebsiella pneumoniae* dinitrogenase reductase
Native molecular weight	225,000	249,800	239,500	66,800
Molybdenum content/atoms	2	0.01	0.03	0
Vanadium content/atoms	0.06	0.085	2 ± 03	0
Iron content/atoms	32 ± 3	24	21 ± 1	3.85
Sulfide ion content/atoms	At least 18	18	19 ± 0.02	4

called ethane (this name is still the approved one). This reduction is the first in the sequence of reactions shown below:

$$HC{\equiv}CH \rightarrow H_2C{=}CH_2 \rightarrow H_3C\text{-}CH_3 \rightarrow H_4C + CH_4$$

The last two transformations to ethane and then to methane (CH_4) are easy for chemists in the laboratory but are generally (but not completely) beyond the capabilities of nitrogenases. The formation of ethene is so ready, and appeared to be so characteristic of nitrogenases, that its production from ethyne was believed to be a specific and unique property of nitrogenases. As such, it was used as a laboratory and field test for nitrogenase activity. It is only recently that it has been shown that there is at least one nitrogenase of a radically different kind that cannot reduce ethyne to ethene.

The ethyne depicted in the reaction sequence above bears more than a formal chemical resemblance to dinitrogen, which can be written N≡N. Indeed, ethyne and dinitrogen are, in chemical parlance, isoelectronic, that is, they possess the same number of electrons. Each carbon atom has four electrons in its outer shell, and each hydrogen atom has one, so ethyne has ten such electrons in all. A nitrogen atom has five outer-shell electrons, so dinitrogen also has ten. The reader can easily confirm this by consulting a standard periodic table. Chemists place great store by the principle that isoelectronic compounds often have similar reactivities. It was only a short step from the reaction sequence above to propose that the reduction pathway of dinitrogen by nitrogenase is as shown below:

$$N{\equiv}N \rightarrow HN{=}NH \rightarrow H_2N\text{-}NH_2 \rightarrow H_3N + NH_3$$

The problem with this proposal, which for a long time was accepted by many biochemists and chemists, is that it is very difficult to demonstrate that it really occurs in vivo. The initial product, HN=NH, the analogue of ethene and called diazene, is a very reactive molecule and difficult to study, its normal lifetime in the chemists' flasks being of the order of milliseconds. The next material, $H_2N\text{-}NH_2$, the analogue of ethane and called hydrazine, is a reasonably stable molecule, even though it is rich enough in energy to be used as a rocket fuel. However, although hydrazine itself may well be related to an intermediate in the reduction of dinitrogen by nitrogenase, it is not a substrate for nitrogenase. In other words, like ethene, nitrogenase enzymes generally do not reduce it any further.[5]

It was finally established that the reduction of dinitrogen to ammonia by the separated proteins under anaerobic conditions may be described by the equation below:

$$16ATP + 8\ e + 8H^+ + N_2 \rightarrow 16ADP + 16P_i + 2NH_3 + H_2$$

To fix nitrogen in vitro in this way, the two proteins are mixed in a buffered (i.e., at controlled acidity or pH) solution in the presence of a reducing agent (a provider of electrons) and the biological energy source (ATP or MgATP if you are being more precise), and the whole mixture is exposed to dinitrogen (figure 6.1). The symbol P_i is used to designate "inorganic" phosphate—that is, phosphate free in solution. This equation apparently represents an optimum efficiency and as it stands is unexpectedly expensive in terms of ATP. In most laboratory situations, the consumption of ATP to reduce a single molecule of dinitrogen is even higher! This has been ascribed to a variety of causes, such as damage to the proteins during the isolation procedures. The big question is: Why is so much energy, two molecules of ATP for each electron transferred to dinitrogen, required to fix nitrogen? Indeed, though not

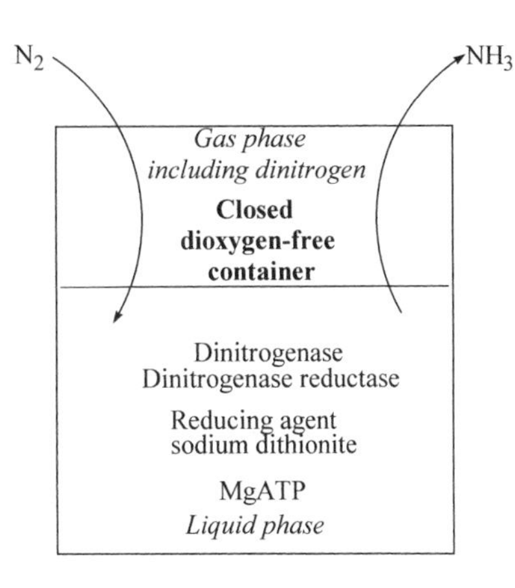

Figure 6.1. How to fix nitrogen in vitro. The protein fractions are generally, but not necessarily, from the same organism. Anaerobic conditions are essential, as well as a controlled physiological acidity level (pH).

widely recognised, there is now at least one very different nitrogenase known that is considerably less demanding in its energy requirements (see below).

It seems that if a nitrogenase system is set up in vitro as described above, with the sole difference that there is no dinitrogen present but solely a completely unreactive gas such as argon, then the system still turns over, but all the electrons arriving at the dinitrogenase protein finally pick up protons (positive hydrogen ions, represented as H^+) from solution and generate dihydrogen, H_2. This reaction is represented below. The single positive charge on a proton is cancelled by the single negative charge on the electron, and the product is a neutral hydrogen atom, two of which then combine to generate dihydrogen:

$$2H^+ + 2e \rightarrow H_2$$

However, the apparently pointless generation of dihydrogen, which occurs even in the most favourable circumstances under optimal fixation conditions when only six electrons of the eight are needed to reduce dinitrogen, still needs an explanation.

At least the large energy consumption can be rationalised to some extent. In terms of energy, the reaction mediated by the cell overall is not running downhill like the Haber–Bosch dihydrogen/dinitrogen reaction, which gives out energy. It is climbing uphill, and it has to be pushed. In formal terms, it can be represented as:

$$\text{Energy} + 2N_2 + 6H_2O \rightarrow 4NH_3 + 3O_2$$

The hydrogen in the ammonia comes ultimately from water, and energy is required to split the water to generate it. Such energy is trapped by the plant as ATP via photosynthesis, and the source of the energy is the sun.

The dihydrogen evolution is so striking that it has even been proposed that a field of legumes, say, of soya beans, could be used both to fix its own nitrogen and to generate dihydrogen for use as a fuel. Nevertheless, each time a molecule of dinitrogen is fixed, six electrons are consumed and at least two are apparently "wasted" as dihydrogen. Why is biology so profligate with hard-won energy? The chemists had an answer to this, as discussed below, but now even that answer looks to be rather suspect.

The availability of the two protein constituents of these nitrogenases enabled biochemists to study in detail how they interact. In 1984, David Lowe and Roger Thorneley at the Unit of Nitrogen Fixation at the University of Sussex developed a model to explain most of the then known facts concerning nitrogenase function (figure 6.2).[6] They proposed that the dinitrogenase reductase receives an electron from a biological reducing agent, a material called a flavodoxin, and that the protein then unites reversibly with the dinitrogenase. Once combined, the electron is transferred to the larger protein. The dinitrogenase reductase and dinitrogenase then split apart and the former then becomes free again, when it can then pick up a further electron from the flavodoxin and repeat the transfer process.

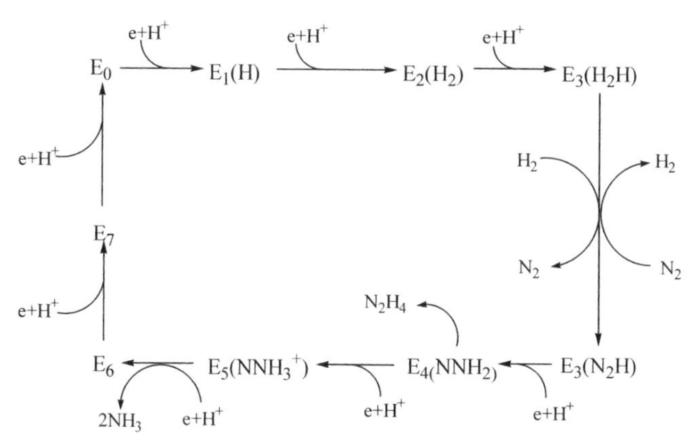

Figure 6.2. The Lowe–Thorneley model for dinitrogen reduction by nitrogenase. The state of dinitrogenase at each of the eight stages in the reduction of a single molecule of N_2 is represented by the symbols E_0, E_1. . .E_7. The arrival of each set (electron plus proton, e + H^+) is shown, as well as the transformations of dinitrogen supposed to occur at each stage. The binding of dinitrogen is supposed to take place only once three protons have been trapped. Two are then displaced (reversibly) by the dinitrogen. This model explains the obligatory generation of dihydrogen and many of the kinetic observations made as the dinitrogen is reduced. The hydrazine may be observed when the whole system is destroyed by quenching with acid at this particular stage of the cycle. Otherwise, ammonia is the sole observed nitrogen-containing product under normal conditions.

Meanwhile, the transferred electron finds its way to the dinitrogen, which is presumably bound somehow to the necessary molybdenum atom, and a proton (positive hydrogen ion) also arrives at the same place to combine with the dinitrogen and maintain neutrality. As indicated above, in formal terms, a positive hydrogen ion plus a negative electron are equivalent to a neutral hydrogen atom. This whole neutralisation process should happen eight times to complete the reaction involving eight electrons and eight protons. In this way, one molecule of dinitrogen is reduced to two ammonia molecules, and a molecule of dihydrogen would escape. So far, so good, but this did not explain what was going on at the atomic level.

The Molecular Biology of Nitrogen Fixation

The study of molecular biology made a significant impact on the problem when R. A. Dixon and J. R. Postgate in 1972 succeeded in transferring a sequence of chromosomal DNA that carried all the molecular information necessary for manufacturing and operating nitrogenase from *Klebsiella pneumoniae* to *Escherichia coli*.[7] This acquisition by *E. coli* of the ability to fix nitrogen had two implications. The first was that since the piece of DNA transferred was relatively small and contained all the genetic information required for nitrogen fixation, it was possible to analyse it and determine precisely what genes were actually involved. Second, it raised the prospect of the transfer of such genetic material to a plant, so the goal of constructing a nitrogen-fixing plant such as of wheat and rice might be ultimately achievable.

The genes, numbering about twenty in all, have now been characterised, and functions have been assigned to most of them. A simplified gene map for *Klebsiella pneumoniae* is shown in figure 6.3. The genes themselves do not carry out the functions associated with them in this figure, but they carry the information necessary to synthesise the proteins and other enzymes that do carry

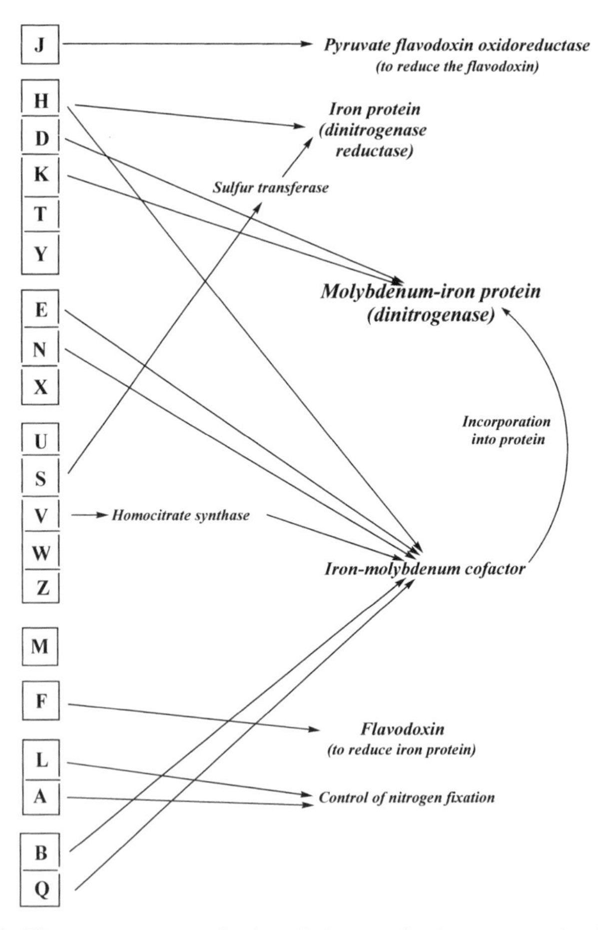

Figure 6.3. The arrangement and roles of nitrogen-fixation genes active in *Klebsiella pneumoniae*. The gene code letters and hence the designations of the individual genes, *nif A*, *nif B*, and so forth, are universally accepted and used, though the letters themselves have no physical significance. The functions of the individual genes, where known, are shown.

out those functions. Subsequent research has shown that all the molybdenum-containing nitrogenases that have been studied (with one notable exception—see below) contain a very similar set of genes, whatever the source, and this initial gene transfer has proved to be the general key to unlocking the genetic basis of nitrogenases in general. In addition, the genes that control the vanadium and iron-only nitrogenases are very similar, though their order and detailed arrangement may differ between organisms. This similarity seems to hold whatever the nitrogen-fixing organism, aerobic or anaerobic, and symbiotic or free-living.

As for transferring the genes to plants, this has turned out to be no trivial problem. To transfer one gene to an alien environment and get it to function is not easy. To get all the genes transferred in such a way that they can all still function together is probably beyond current science. For instance, a green plant undergoes photosynthesis, a function that produces dioxygen, which, in turn, is usually a poison for nitrogenases. If the nitrogen-fixing gene complex were placed in a cell of the leaf of a green plant, some arrangement, such as is found in many natural aerobic nitrogen–fixing systems, would have to be made to protect the enzyme from the poison. The construction of such a protective system from scratch where none exists will be very difficult.

The next great leap forward was in the early 1980s. Paul Bishop showed that under certain circumstances of molybdenum starvation (and this is difficult to achieve because many organisms are highly efficient at scavenging molybdenum even when only tiny traces are present) *Azotobacter vinelandii* could still fix nitrogen, but evidently using a nitrogenase that does not contain molybdenum.[8] This discovery was initially greeted with scepticism. Indeed, one authority wrote in 1982 that: "Reports that an active vanadium nitrogenase can exist are probably false." Nevertheless, it will be remembered that Bortels fifty years earlier had shown that vanadium could support nitrogen fixation, and vanadium finally turned out to be the necessary metal contained in this new nitrogenase. This was ultimately proved genetically by excising the genes of *A. vinelandii* responsible for the processing of molybdenum and finding that the engineered organism could still fix nitrogen. Genes for processing vanadium were subsequently uncovered.

One can go a stage further and excise both the molybdenum- and vanadium-processing genes and still obtain a functional nitrogen-fixing *A. vinelandii* strain. This indicates yet a third kind of nitrogenase. This kind of nitrogenase seems to contain iron only and differs slightly in its properties from the two others, which are very similar. In fact, it is generally believed that the only significant structural difference between the active sites of the molybdenum and vanadium nitrogenases is that vanadium in the second takes up the position occupied by molybdenum in the first. Otherwise, their structures are reckoned to be the same and their functions very similar. The compositions of selected kinds of dinitrogenase are summarised in table 6.1.

The relationship of these three nitrogenases in an organism that is capable of producing all of them is intriguing. It seems that the molybdenum nitrogenase is always the preferred variety. If the organism is not to die in the complete absence of molybdenum and fixed nitrogen from its environment, and if it has the ability to produce the vanadium nitrogenase, it will fix nitrogen using that enzyme. If vanadium is also completely absent, those lucky organisms that can do so will fix nitrogen using the iron-only nitrogenase. Nature seems to have evolved a graduated response to providing fixed nitrogen, though if fixed nitrogen is available in a sufficient amount then the appropriate genes may never be switched on and no nitrogenase of any kind produced. Presumably, this is an energy-saving measure.

The Structure of Nitrogenase and the Mechanism of Nitrogen Fixation

The method used most often for determining the positions of atoms in a molecule is X-ray diffraction, and from the time that the separate nitrogenase proteins were isolated, scientists have attempted to apply this method to determine their structures. The problem is that, until very recently, doing so has always required that the substance under investigation be crystalline and, more than that, be present as a single crystal. The preparation of single crystals is relatively easy with simple materials such as common salt, sodium chloride, but with proteins it can be very difficult. Consequently, it was thirty years after the first preparation of cell-free extracts of a nitrogenase that a structure was finally determined. In 1992, the structure of the larger molybdenum–iron protein from *A. vinelandii* was first described.[9] Here surely would be the answer to all our questions about the details of the fixation at an atomic level, for all the atoms involved were to be revealed.

The result was completely unexpected. The molybdenum, presumed to be the seat of all the action, was shown to be part of an assembly of iron and sulfur atoms, of a type known as a cluster. In the particular case studied, the cluster contains molybdenum, iron, and sulfur, and it is actually extractable from the protein as a single unit. It is sometimes called the iron–molybdenum cofactor, or FeMoco. This is shown in figure 6.4, which portrays the delicate network of alternating iron and sulfur atoms that constitutes the major part of the cluster. The structure is best imagined as a kind of sausage that is tied at either end to the supporting protein. This in itself is highly unusual, and there is no precedent for this in any other metalloenzyme. One end of the sausage is an iron atom, and the other is a molybdenum atom, for so long believed to be the site of the mysterious nitrogen-fixing ability. The sausage should not be imagined as swinging like a hammock in a void. Actually, it is surrounded by ordered water molecules that must play some part in delivering the protons necessary to balance the electrons used in the reduction of dinitrogen. However, most of the iron atoms seemed rather exposed in terms of the concepts usually used by chemists. They are each joined only to three other atoms, whereas at least four might have been expected. Does this mean that they are capable of binding a further group, such as dinitrogen or ethyne? In addition, the molybdenum has six nearest neighbours, which is a number often

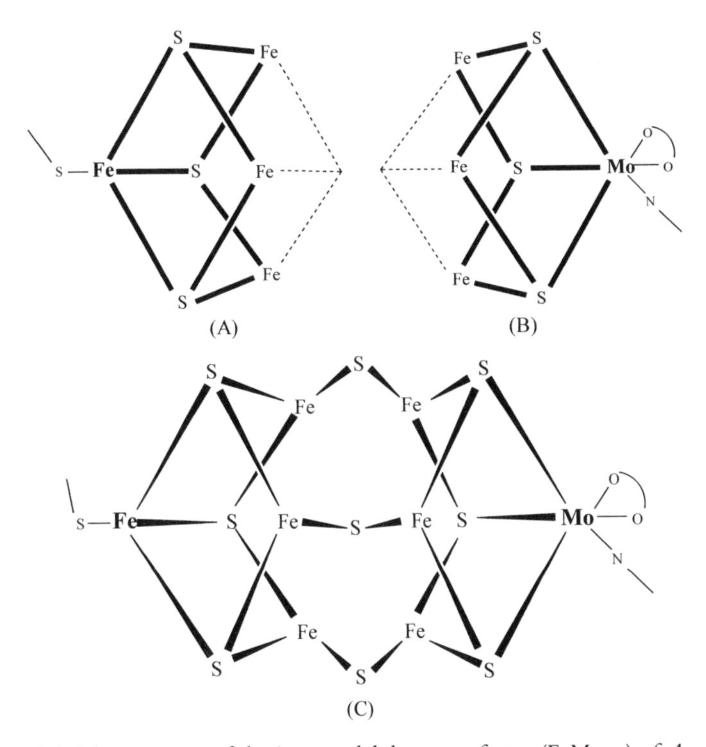

Figure 6.4. The structure of the iron–molybdenum cofactor (FeMoco) of *A. vinelandii* as revealed by the initial studies of Rees et al., elaborated by Bolin et al., and confirmed in many other researches. The structure may be regarded as the association of two cubic arrangements of metal and sulfur atoms, (A) and (B), of which one corner is empty. These two cubic residues are united by bridging sulfur atoms between pairs of corresponding iron atoms (C). The unique iron atom is bound to the protein by a cysteinyl residue (sulfur shown), and the molybdenum atom is bound to the protein by a histidinyl residue (nitrogen shown). The two oxygen atoms bound to the molybdenum are part of a single homocitrate molecule.

associated with molybdenum in stable (saturated) complexes. One of the compounds involved in binding to molybdenum is the anion of a rather unusual organic acid, homocitric acid. Why it should be there is not obvious. However, if the molybdenum is really completely tied up, how can it combine with dinitrogen, which must have an open site at which to bind? Does molybdenum have anything at all to do with the actual trapping of dinitrogen?

The structure reignited the question of how the enzyme works at the atomic level. Since 1992, the structures of several different molybdenum nitrogenase enzymes (for example, of *Clostridium pasteurianum*) have been determined, and they are all very similar to that of *A. vinelandii*. Even structures of the two nitrogenase proteins bound together as they exchange an electron between them have been unravelled. The data all concur concerning the shape and constitution of the molybdenum-containing cluster.

These structures can be considered to be pictures of the enzyme system but taken at a specific point during its turnover. However, we already know that eight electrons are passed into the system during the reduction cycle, and presumably the structure changes to some degree as each electron arrives at the reduction site. There cannot be a single structure but eight related structures, all of them equally valid representations of the enzyme. To what particular stage of the reduction cycle, if any, do the structures determined actually correspond? Unfortunately, no one has been able to take such a picture of the enzyme with dinitrogen or any other substrate actually bound to the active site, and even indirect evidence, such as might be obtained by spectroscopy, is very sparse. If the molybdenum cannot bind dinitrogen, then presumably the iron does so. However, is the structure determined by X-ray analysis of the crystalline protein really the way that the atoms are arranged within solution in a cell? The mere act of crystallisation may impose an artificial structure upon the protein and the cluster. Consequently, we still cannot exclude the possibility that molybdenum is at the active site. On the other hand, if the active site is iron, then is it a single iron atom, or is it some combination of the iron atoms? If it is a combination of iron atoms (or even of iron and molybdenum), does the dinitrogen stick initially to the inside or outside of the cluster? If the dinitrogen really goes inside the cluster, how and where does it enter, how do the necessary protons enter and attack the dinitrogen, and how does the product ammonia get out? All these questions and many more have been addressed in numerous publications, and different authorities favour different answers. The determination of the enzyme structure, which, it had been fondly hoped, would reveal the ultimate secret, has served only to show us that there is another layer of problems yet to be solved.

Theoretical chemists got involved in all this, too, because the idea that dinitrogen could interact with a metal–sulfur cluster such as the iron–molybdenum cofactor was very novel. Calculations were made to justify a whole range of proposals as to how that interaction might proceed, even given that there was (and still is) no direct evidence that a simple interaction of dinitrogen with the cluster in the form and oxidation level in which the enzyme crystal is prepared ever occurs. Interaction at the molybdenum would require some rearrangement from the observed structure, but the dinitrogen might conceivably stick to an iron atom, or between two iron atoms, or it might even enter the interior of the cluster. It may even occupy different positions at different stages of the reduction cycle. Some of these possibilities are indicated in figure 6.5.

Since the seminal paper of Jongsun Kim and Douglas Rees, the structure has been refined and made more precise. Slowly, more detail has emerged, but in 2002 Rees and his collaborators provoked another agonising reappraisal.[10] At the highest resolution yet achieved, a shadowy atom, perhaps carbon or even nitrogen, made its appearance at the heart of the cluster (figure 6.6). Whatever

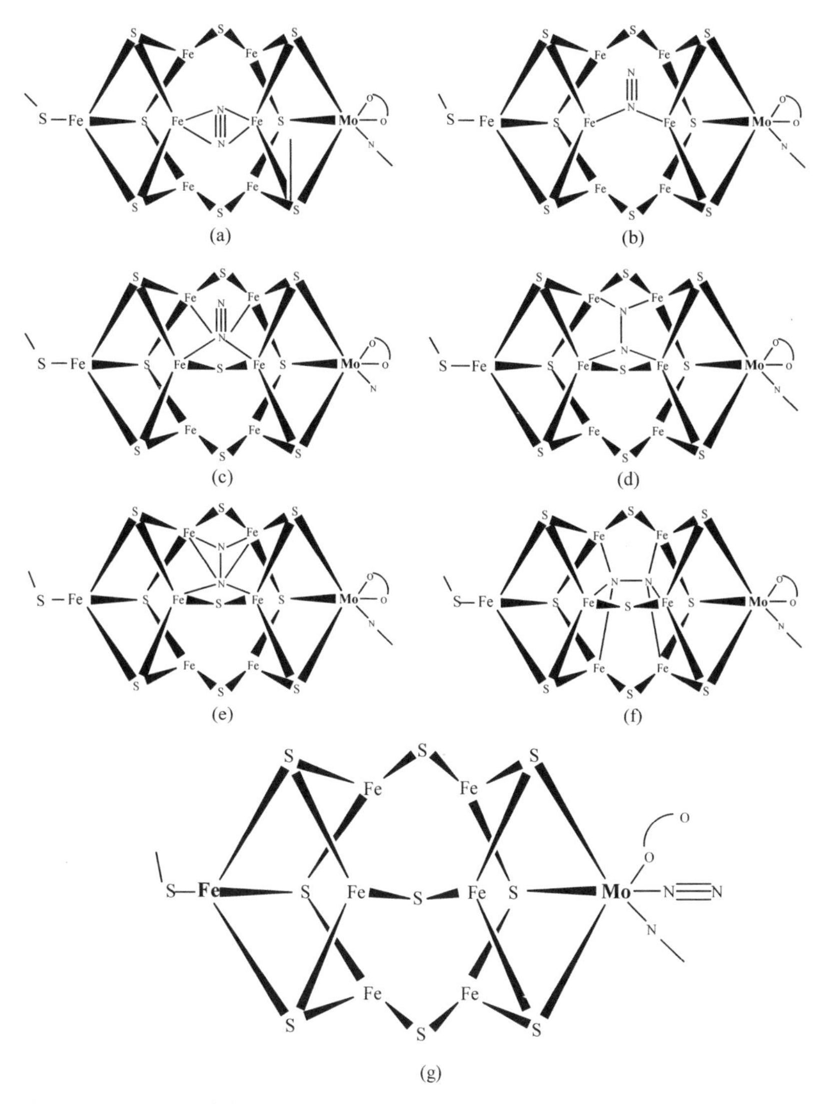

Figure 6.5. Some of the possibilities for the interaction of dinitrogen with the iron–molybdenum–sulfur cluster at the heart of dinitrogen reductase. (a) shows a dinitrogen bound between two iron atoms, having displaced a sulfur; (b) is a variant on this; (c) shows dinitrogen binding end-on to four iron atoms and (d) is similar but with both nitrogen atoms involved; (e) is an asymmetric version of (d) and (f) involves the dinitrogen and all six "interior" iron atoms; finally, (g) shows binding of dinitrogen to molybdenum, which also carries the histidine (indicated by -N-), but where the homocitrate, indicated by the oxygen atoms, has lost one of its two bonds to molybdenum, leaving a place open for dinitrogen to bind.

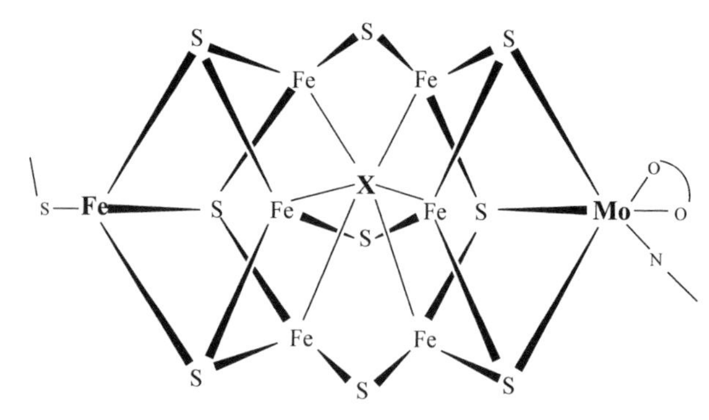

Figure 6.6. The structure of the cluster at the heart of *Azotobacter vinelandii* dinitrogenase as of September 2002. An "extra" light atom, perhaps carbon, nitrogen, or oxygen, was discovered in high-resolution X-ray structural studies to be at the centre of the iron–molybdenum cofactor. Its significance is far from clear.

it is, the presence of this atom seems to show that there is unlikely to be any dinitrogen chemistry within the cluster. If the interaction is outside the cluster, it has to be accepted that the iron atoms do not now appear to be in such an unusual environment as has seemed to be the case hitherto because they can also bond to the newly identified light atom. If so, then is the molybdenum again the more likely candidate as the site of dinitrogen binding? If it is, then what happens to it to make it more likely to bind dinitrogen since it seems unlikely to do so if it remains in the state detected in the crystals? The opportunities for experiment and speculation seem boundless. The more we know, the less we understand.

The Chemistry of Biological Nitrogen Fixation

None of the discussion above throws light on a more general chemical conundrum: How does the unreactive molecule dinitrogen react with other compounds under mild conditions (such as pertain inside an enzyme) when industrial chemists have found it necessary to use extremes of temperature and pressure to make this happen? At about the time that the first cell-free extracts were obtained, in 1960, chemists were also beginning to make fundamental advances in the chemistry of nitrogen fixation. Haber–Bosch-type catalysts functioned in a way that was reasonably well-understood even then, the surface of the metallic iron effectively generating a sea of individual nitrogen atoms and hydrogen atoms by splitting the molecules of the reactants, dihydrogen and dinitrogen. Eventually, some come together to form ammonia, and this then boils off the catalyst surface. However, the way in which a single metal atom rather than a catalytic metallic surface composed of a multitude of such atoms might mediate the formation of ammonia, and under mild conditions, was completely unknown.

There certainly were ideas about how this might happen, at least in the initial stages. For example, carbon monoxide (CO) and dinitrogen are, in chemists' terms, isoelectronic. They might be expected to behave similarly in some respects. The analogous parallel between the triple bond in ethyne and dinitrogen has already been discussed. Carbon monoxide had been known to bind to metal atoms for at least seventy years. In fact, Brunner, Mond & Co., the company that grew into ICI, might be said to have been founded on one such compound, nickel tetra(carbonyl). If the letter M represents a typical metal atom, then the basic structural motif of such a material may be represented by the set of symbols M←C≡O. If carbon monoxide could do this, why couldn't dinitrogen, to give M←N≡N? On the other hand, since ethyne was also known to bind to metal atoms, but in a different sense,

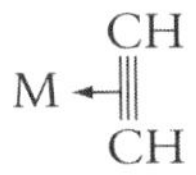

$$M \leftarrow \begin{array}{c} CH \\ ||| \\ CH \end{array}$$

should not dinitrogen be able to do something similar and form compounds with an analogous structure?

$$M \leftarrow \begin{array}{c} N \\ ||| \\ N \end{array}$$

Despite such speculations, no chemical evidence for either of these possibilities was forthcoming. However, nitrogen fixation became a topic of research of consuming interest, and, more important, in the 1960s government authorities were prepared to fund it quite generously. Chemists in the United States, the Soviet Union, and Japan were hot on the trail of the elusive trap for dinitrogen. The United Kingdom set up a well-found laboratory called The Unit of Nitrogen Fixation at the University of Sussex, with the remit of determining the mechanism of biological nitrogen fixation. This Unit was multidisciplinary. Because the chemistry of nitrogen fixation seemed at that time to be so unusual, members of the Unit included inorganic chemists as well as the more usual kinds of biologist and biochemist. Nevertheless, the initial chemical advances were made elsewhere.

The first hint of success in chemistry came in 1964, with the publication by Mikhail Vol'pin and his group in Moscow of the discovery that if one mixed a salt such as iron(II) chloride with a very strong reducing agent in the presence of dinitrogen, then the gas was absorbed and reacted.[11] Subsequent treatment of the resulting mixture with aqueous acid produced ammonia. Some believe that the dinitrogen is actually split by the mixture to form negatively charged nitrogen atoms, or nitride ions, perhaps not unlike the action of a Haber–Bosch catalyst, but the detailed reaction mechanism is not clear even today. Nevertheless, this constituted the first generally recognised sign of opening the chemistry box of dinitrogen secrets.

The next major advance followed soon after. In 1965, Allen and Senoff in Toronto described the first compound in which dinitrogen was bound directly to a metal ion, in this case ruthenium, which has the symbol Ru.[12] The compound contained the structural element $Ru \leftarrow N \equiv N$, and in a way that was common at the time, the discovery was accidental. The senior author, A. D. Allen, had a lot of trouble convincing the chemical community that he had indeed isolated such a compound. It was only as chemical understanding grew that more rational syntheses were devised. Nevertheless, it remains generally true that none of the dinitrogen compounds prepared then or subsequently is conveniently obtained from air. Dioxygen from the air invariably destroys the compounds involved in trapping dinitrogen, and even water often does so. The model complexes that take up dinitrogen and the nitrogenase enzymes are similarly sensitive to dioxygen. The goal of cleanly absorbing dinitrogen directly from the air to generate a pure dinitrogen complex has not yet been attained.

Dinitrogen complexes soon began to appear in a stream that became a flood. Shortly afterward, in 1966 and 1967, various workers in Japan, Italy, and the United States described iridium–dinitrogen and cobalt–dinitrogen complexes.[13] A complex with dinitrogen bound between two ruthenium atoms and containing the grouping $Ru \leftarrow N \equiv N \rightarrow Ru$, was reported in 1968.[14] This was taken to be a precedent for dinitrogen to bridge between metal atoms, perhaps as between molybdenum and iron in dinitrogenase. It was found that the Allen and Senoff compound could actually be prepared in aqueous solution in a reaction in which dinitrogen displaces water from its combination with the metal atom. Such a reaction could parallel what goes on in the aqueous solution inside a cell. Dinitrogen is now known to be able to bind to a single metal atom, as discussed above, or to bind to two, three, or four metal atoms simultaneously. Probably several hundred dinitrogen complexes are known, with dinitrogen bound to a large variety of metal atoms. When they were first announced, the relevance of most of these compounds to the study of biological nitrogen fixation seemed (and still seems) to be marginal. Nevertheless, the belief persisted that if enough was discovered about the chemistry of dinitrogen, then together with the structure of nitrogenase, we would be able to point to the behaviour of dinitrogen in an apparently analogous situation and quickly discern how the dinitrogen would react in the enzyme.

Two developments encouraged this belief. One was the observation in 1967[15] that dinitrogen could displace dihydrogen from combination with a metal (initially that metal was cobalt, but other examples were soon found). This is shown by the equation below.

$$M\begin{matrix} \diagup H \\ \diagdown H \end{matrix} + N_2 \longrightarrow M \leftarrow N \equiv N + H_2$$

Here surely is the explanation of why dihydrogen is always produced by nitrogenases. It is necessary to displace at least one molecule of dihydrogen for every molecule of dinitrogen that is reduced. The suggestion that hydrides might be involved in the function of a metalloenzyme was then novel but not completely unreasonable. It has already been discussed above in the context of the Lowe–Thorneley model.

The second observation to reinforce the hope that the mechanism of biological nitrogen fixation was close to being uncovered was the announcement in 1969 that molybdenum could form a dinitrogen complex, and this was the first of a relatively large number of molybdenum–dinitrogen complexes.[16] By 1970, it was possible to claim that the puzzle of how nitrogenase fixed nitrogen was almost solved. All that was necessary was to make dinitrogen in these model compounds react somehow to generate ammonia.

As already indicated, dinitrogen complexes have now been discovered in which dinitrogen is bound to a metal in a variety of ways to one, two, or more metal atoms. When this happens, the nitrogen–nitrogen link of the dinitrogen is stretched and weakened, so that in formal terms one can infer that the nitrogen–nitrogen triple bond can become a double bond, then a single bond, and finally no bond at all. That individual molybdenum atoms in particular complexes can actually do this unaided was determined in 1997 in a reaction that is summarised by the equation below[17]:

$$\mathrm{Mo} \leftarrow \mathrm{N{\equiv}N} \rightarrow \mathrm{Mo} \longrightarrow \mathrm{Mo{\equiv}N} + \mathrm{N{\equiv}Mo}$$

Since then, another biological metal, vanadium, and also niobium have been found to mediate a similar reaction, and there is even an example where two different metal types can combine at either end of a dinitrogen molecule to do this. However, the established structure of dinitrogenase (see above) makes it very unlikely that any of these reactions is involved in nitrogenase function. It is surprising that in this wealth of new chemistry, an unequivocal example of a stable complex with dinitrogen bound in the manner typical of ethyne, side-on to a single metal, has yet to be discovered.

During the 1970s and 1980s, the reactivity of bound dinitrogen slowly became evident. Some complexes, for example of titanium, were found to produce hydrazine, H_2NNH_2, upon treatment with acid. Other systems, some based upon molybdenum, were shown to be able to reduce dinitrogen to hydrazine and/or ammonia catalytically in aqueous solutions. It should be noted that none of these fixation systems was stable in the presence of dioxygen, either pure or in air.

The most encouraging system to be developed in this period was based upon dinitrogen complexes of molybdenum or of the chemically related metal tungsten. It was shown that such a metal dinitrogen complex could be converted to ammonia by treatment with acids in a series of reactions shown by the equation below[18]:

$$\mathrm{Mo}^{0} \leftarrow \mathrm{N{\equiv}N} + 6\mathrm{H}^{+} \longrightarrow \mathrm{Mo}^{\mathrm{VI}} + 2\mathrm{NH}_3$$

In this equation, the six electrons required to convert the dinitrogen to ammonia all come from the metal, in this case molybdenum, in a very reduced form. (The superscript 0 in the equation above indicates a very reduced form of the metal, oxidation state zero, and the superscript VI in the equation indicates that the metal has become oxidised to oxidation state six.) Treatment with acid provides the protons, and the metal loses the electrons to become oxidised. The parallel with the enzymatic dinitrogen-protonation reaction is very clear:

$$N_2 + 6H^+ + 6e \rightarrow 2NH_3$$

Further detailed research showed that these six electrons are transferred from the molybdenum to the dinitrogen stepwise and that the system passes through a range of intermediate stages until the dinitrogen conversion to ammonia is complete.

It was an extension of this observed chemistry that led to a proposal for the reduction of dinitrogen bound to molybdenum in dinitrogenase, and one form is shown in figure 6.7. The six electrons formally required to reduce a molecule of dinitrogen to ammonia were shown to reach it one at a time, as protons were picked up from solution to maintain neutrality. This is repre-

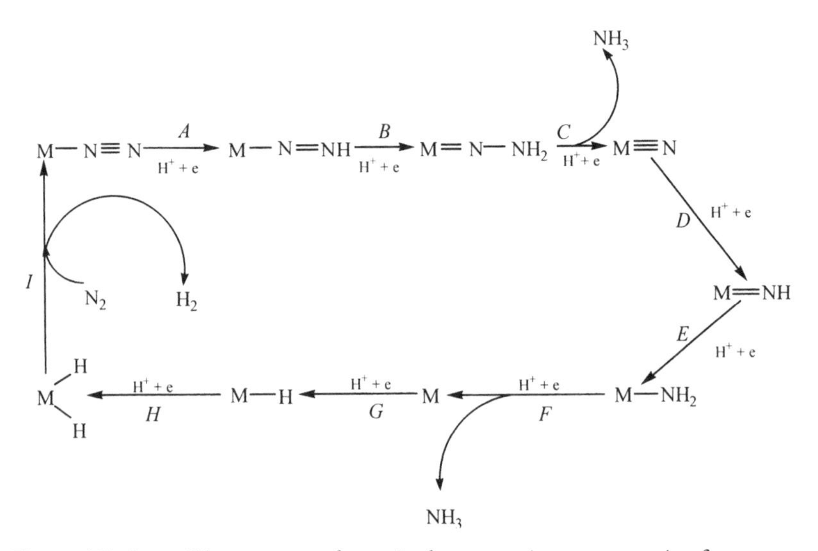

Figure 6.7. A possible sequence of steps in the conversion to ammonia of dinitrogen coordinated with molybdenum. This cycle, based upon established dinitrogen and molybdenum chemistry, may provide a rationale for the observed reduction of dinitrogen both in model systems and by molybdenum-containing nitrogenases. If this really occurs in natural systems, then the oxidation state of the molybdenum is unlikely to drop as low as zero, but such systems have means of storing electrons other than simple reduction of the key metal atom.

sented in steps A–F in figure 6.7. Steps G-H in the figure involve supplying two more electrons from an external source and balancing their negative charges with two more protons from solution. This produces a metal dihydride, and there is plenty of precedent for reactions of this kind. The final step to close the cycle, step I in figure 6.7, is the displacement of dihydrogen by dinitrogen, for which, as already mentioned, there is also a precedent. This proposal explains why eight electrons are necessary for the complete catalytic cycle in the enzyme and why dihydrogen is inevitably generated. What is unlikely is that this kind of transformation could be mediated by a simple single-metal complex because no such compound would be expected to retain its basic structure if it were to lose six electrons.

All the stages in this cycle are represented in the related model chemistry by stable compounds, most of which have been isolated from dinitrogen-complex-protonation reaction mixtures, and though the simplest model systems are completely destroyed upon exposure to acid when the ammonia is generated, the cycle can be turned successfully and repeatedly using the techniques of electrochemistry.[19] Presumably, nitrogenase can also do this by virtue of being able to store and release electrons as required. Very recently, a related system has been described[20] that produces ammonia from a molybdenum–dinitrogen system that cycles between oxidation states III and VI. These oxidation states are much more likely to be produced in a biological system than the 0 and II states involved in the earlier purely chemical variant.

These are the most complete descriptions of an empirical conversion of a coordinated dinitrogen molecule to ammonia. Though aqueous systems based upon molybdenum are capable of converting dinitrogen to ammonia catalytically, under mild conditions, and at pressures of dinitrogen of one atmosphere, the detailed reaction mechanisms have not been fully elucidated, so a comparison with the data upon which figure 6.7 is based is just not possible.

Surely, if Nature really knows her chemistry, she must do something similar to the cycle shown in figure 6.7? However, it is very possible that she doesn't do things this way at all. The elucidation of the structure of the presumed active site of dinitrogenase has raised considerable doubts as to whether this elegant cycle has any relevance to the function of nitrogenase enzymes. It may, if the structure of the nitrogenase cluster containing the molybdenum changes significantly when the enzyme is turning over. In that case, the structure observed in the crystal may alter to allow dinitrogen to bind to molybdenum and allow the subsequent reactions depicted above to take place. There is not much evidence for or against this view at this point, despite the efforts of many researchers over more than ten years. A recent suggestion[21] attempts to explain why the very unusual compound homocitrate is bound to molybdenum and how it actually rearranges during turnover to allow dinitrogen to bind. What is sure is that chemistry of biological nitrogen fixation is almost as much a mystery as ever it was.

The Twist in the Tail

It seems that the burst of information that appeared in the 1960s and the subsequent decades has slowed to a trickle. In 1987, *The Times* of London was making optimistic reports about the possibility of making nitrogen-fixing plants, but today one can no longer whisper the magic words "nitrogen fixation" to granting bodies and expect to be funded. The Unit of Nitrogen Fixation, at one time the leading laboratory in the world in this area of research, has disappeared. After forty years of listening to researchers who honestly believed that they were on the verge of the final breakthrough, the economic advantages of most of this very elegant and detailed work are not evident to funding bodies. That is not to say that nothing of benefit has accrued. For example, work with symbiotic systems involving rhizobia has indeed found application, but the ultimate mystery of biological nitrogen fixation is still to be solved. The way in which genes are controlled has been beautifully illuminated by much of the nitrogen-fixation research. However, in 1997, Nature provided another twist in the story, the significance of which has yet to be fully understood.

All the nitrogenases discussed so far are very dioxygen-sensitive. They generally operate in organisms living in the soil or in fields, on leaves or roots, at temperatures within the normal ambient ranges. However, in 1997, Ortwin Meyer and his collaborators[22] described a nitrogen-fixing organism that they found in the wood and turf walls of a traditional charcoal-producing furnace. This form of charcoal production is the conventional way of making it and has been practised in many different countries for some hundreds of years, though it is now generally obsolete. The organism, *Streptomyces thermoautrophicus*, lives in the walls of such furnaces and fixes nitrogen happily at 60 °C. Its nitrogenase contains both iron and molybdenum, but it consists of three proteins, one of which, a dinitrogenase, contains both iron and molybdenum. However, it differs from the dinitrogenases discussed above (table 6.2). The whole system requires dioxygen to function, whereas dioxygen needs to be excluded in the "conventional" or "classical" cases. It converts carbon monoxide to carbon dioxide as it turns over, though carbon monoxide is a poison to the "conventional" nitrogenases. This new nitrogenase cannot reduce ethyne, whereas such a capability is characteristic of all the other nitrogenases. It produces dihydrogen but less than the other nitrogenases, and, unlike them, dihydrogen does not inhibit its function. Apparently, it can use dihydrogen as an energy source. It consumes ATP at a lesser rate than conventional nitrogenases, so it seems to be more efficient. An approximate equation for the reduction of dinitrogen in this system is shown below.

$$N_2 + 4\text{–}12ATP + 8H^+ + 8e \rightarrow 2NH_3 + H_2 + 4\text{–}12ADP + 4\text{–}12P_i$$

Apparently, one molecule of dihydrogen is produced initially for each molecule of dinitrogen reduced, but in the best circumstances only the equivalent of one-half of a molecule of ATP is needed to generate each electron as

Table 6.2

A comparison of the two species of molybdenum nitrogenase. One contains two constituent proteins and the other three. The latter is the nitrogenase from *Streptomyces thermoautropicus* and is not air-sensitive, whereas the former is. The latter could never be detected using the ethyne-reduction test. The existence of these two types of nitrogenase hints that there may be other as yet undetected vatieties of nitrogenase adapted to further environments.

	Protein constituents	Molecular weights	Molybdenum content	Iron content	Sulfide content	Other significant metals	Dioxygen sensitivity of enzyme	Ethyne reduction by enzyme
Nitrogenase of *Klebsiella pneumoniae*	Dinitrogenase	225,000	2	32 ± 3	At least 18		Great	Characteristic of all such enzymes
	Dinitrogenase reductase	66,800	0	3.85	4			
Nitrogenase of *Streptomyces thermoautrophicus*	St3 (Carbon monoxide dehydrogenase)	273,000	0.77 ± 0.1	7.54	25	Zinc, 0.39, copper, 0.8, no vanadium	Dioxygen is a requirement for nitrogen fixation	Cannot reduce ethyne
	St2 (Superoxide oxidoreductase)	48,000	<0.04	0.07	0	Zinc, 0.1, manganese, 0.25, no vanadium		
	St1 (Dinitrogenase)	144,000	0.7 ± 0.1	12 ± 1	12 ± 1	Zinc, 3.4, no vanadium		

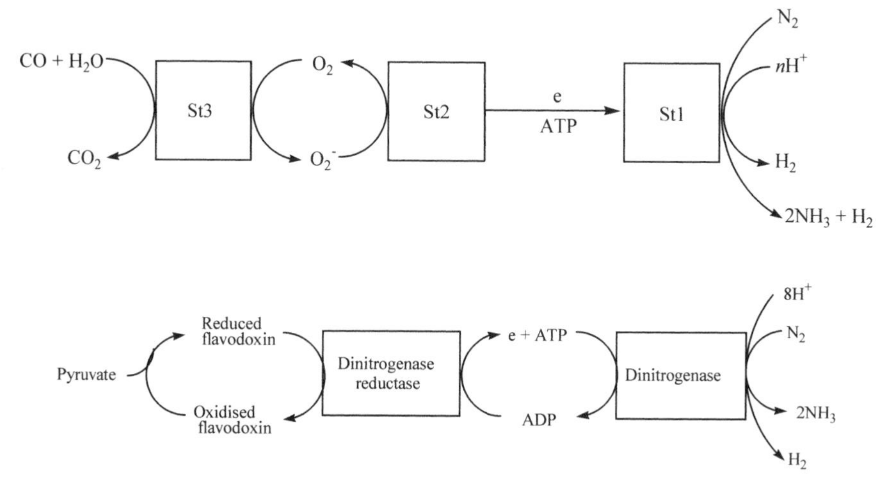

Figure 6.8. The modes of operation of the two types of molybdenum nitrogenase. The "new" nitrogenase requires not only dioxygen, from which the "classical" nitrogenases have to be shielded, but also carbon monoxide, which inhibits nitrogen fixation by the "classical" enzymes.

compared with two in the "conventional" systems. What these properties imply for the detailed function of the new nitrogenase is not clear. However, at this stage, it appears that there is a new kind of nitrogenase that is quite different from those we have so far recognised. Its function is sketched in figure 6.8, which also shows a comparable diagram for a "classical nitrogenase." This raises a host of further questions. Are there as yet other unidentified environments that hide nitrogen-fixing organisms? Have we therefore underestimated the magnitude of biological nitrogen fixation because we have not yet counted all the sources? Do these as yet unidentified organisms fix nitrogen in what has been regarded hitherto as the conventional way, or does Nature have more ways of doing this than we know? Certainly, the known laboratory chemistry of dinitrogen is now very rich and some of it may be exploited in as yet unsuspected ways. Nevertheless, the ultimate secret of biological nitrogen fixation still eludes all attempts to uncover it. The feasibility of producing engineered nitrogen-fixing plants has still to be confirmed, and it is unlikely that this will happen soon. There is evidently still much to learn.

CHAPTER 7

Nitrogen, Threat or Benefaction?

The Spectre at the Feast

Nitrogen at the Beginning of the Twenty-First Century

The world today is a very different place from what it was in about 1900. It is a very different place from what it was even in the 1960s. This is not to say that the worries and preoccupations of 1900 and the 1960s have just disappeared. Rather, they still remain, but as a consequence of the activities of the Club of Rome and the many similar organisations that have arisen since then, people are much more conscious of them. The famous energy crisis of 1973, provoked by the rapid quadrupling of the price of oil, hardly a natural process, served to push such considerations to the fore.

The simple questions that were once posed (such as "How shall we feed a growing population?") have been joined to many others. Is there a limit to population growth beyond which the potential food supply will really be exceeded? Is there a limit beyond which the perturbation of the environment by human actions will produce changes that will irretrievably damage both people and the environment? Are there really limits to growth? What can we reasonably do that will not produce disaster? This is a far cry from the Victorian and even old-fashioned capitalistic and Soviet attitudes that seemed then and still seem to assume that humans, being at the pinnacle of evolution (or, alternatively, being placed at the pinnacle of animal life by God), were free to exploit Earth and its resources as much as seemed necessary.

Even to attempt to answer such questions, it is necessary to understand what the current state of Earth and the environment really are, and this is not simply a matter of looking out of the window and making a snap judgement, or even looking out of several windows over a certain period. It is necessary to do serious research and then attempt to make sound judgements. This is no trivial matter because often there is little objective guidance as to what constitutes a sound judgement.

Humans and the Environment

The idea that human activities are upsetting the current equilibrium between people and the environment is based upon a misconception. There is not and never has been a significant equilibrium in the environment. This was true before humans or any growing things appeared upon the scene, and the rate of change may not necessarily have been slow. This is not to say that human activities are without effect or without significance. However, until one can assess the changes on Earth that might be occurring due to influences such as variation in the emission of radiation by the sun or in the inclination of Earth's axis, or the mean differences in pressure over the Atlantic Ocean between Iceland and the Azores, let alone normal geological processes taking place on Earth's surface, one cannot really be certain of the influence humans are having on the environment. For example, global warming and cooling have been constant features of geological history since long before humans arrived upon the scene (figure 7.1).[1] The overall range in the diagram shown is about 40 °C. Data going back millions of years can be estimated through isotope measurements on ice and mud cores, but the data since about 1860 are based upon direct instrumental measurements. These are easier to understand and are shown in figure 7.2.[2] Earth is getting warmer, but it is not altogether clear why.

Nevertheless, it is common knowledge that humans have changed the face of Earth in the past. Many of the forests of Western Europe have disappeared as people became settled farmers. One of the first British (English) energy crises hit in the sixteenth and seventeenth centuries when the supply of trees that provided the ships of Henry VIII and Elizabeth I ("Hearts of oak are our ships") as well as the charcoal that was then the basis of the iron industry became less available. The country was simply running out of suitable trees. In the case of the iron industry, as so often happens, another source of fuel, coal, was found. Nevertheless, the environment was changed irreversibly.

However, the changes of climate occurring in the recent past have been far from trivial. Desertification, as in the Sahara Desert, is not a new phenomenon. Humans lived there in an apparently clement environment not so many years ago. The River Thames at London froze over regularly during the Middle Ages but never does so now. The list of such changes is very extensive. Some observers feel sure that these changes can be ascribed to human activities. We know that humans felled trees in what is now the Sahara Desert, and that the River Thames now runs in a very narrow and restricted bed compared to the floodplain of some 400 years ago. Nevertheless, it is quite evident (for example, from appropriate isotope measurements in ice cores and

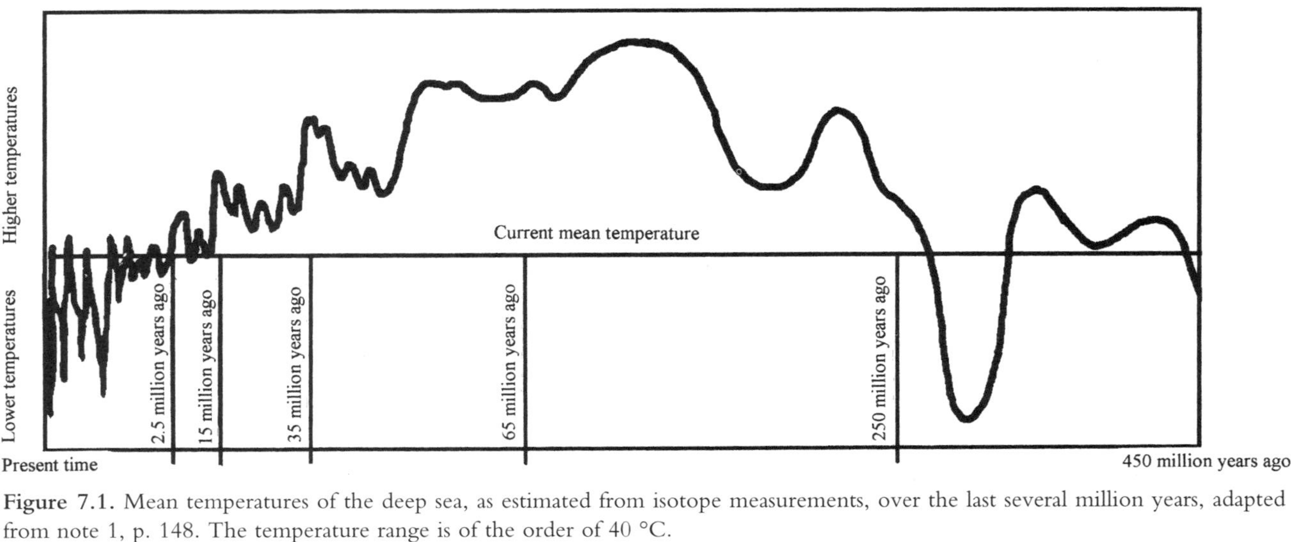

Figure 7.1. Mean temperatures of the deep sea, as estimated from isotope measurements, over the last several million years, adapted from note 1, p. 148. The temperature range is of the order of 40 °C.

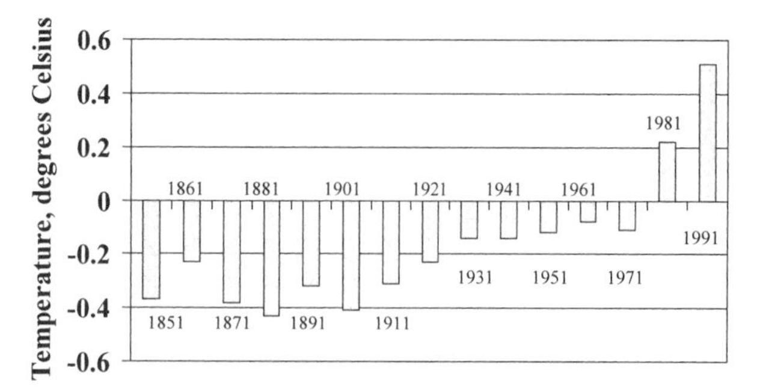

Figure 7.2. Recent temperature variations in the northern hemisphere. The temperatures are ten-year averages for the decades beginning in the years indicated, as a departure form the average for 1961–1990. For a detailed discussion, see P. D. Jones and A. Moberg, *J. Climate*, 16, 206 (2003).

elsewhere) that not only have mean ambient temperatures changed significantly over the millennia but that levels of carbon dioxide in the atmosphere have been many, many times greater than the levels observed today (figure 7.3).[3] The increase observed in modern times (figure 7.4)[4] is relatively minor, but worrying because we cannot yet know all the consequences nevertheless, even if it turns out not to be due to human activities.

During the millions of years of Earth's history, the dioxygen content of the atmosphere was sometimes both much more and considerably less (probably even zero before green plants developed on land) than it is today. The high levels of dioxygen do not seem to have provoked spontaneous combustion of vegetable matter, though thermodynamic calculations seem to suggest that they should have done so. Despite such elevated carbon dioxide and dioxygen levels, there is no evidence that the world overheated due to an enormous greenhouse effect, and vegetation was not completely consumed by fire. Other factors must have intervened.

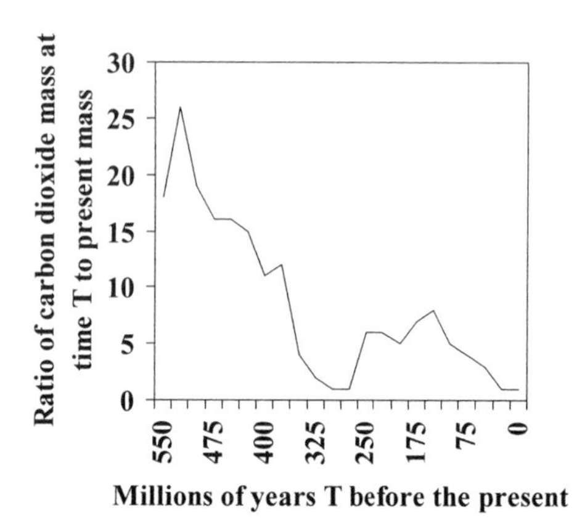

Figure 7.3. Estimated historical levels of carbon dioxide in the atmosphere. Data were supplied by Professor R. A. Berner. For a complete discussion, see R. A. Berner and Z. Kothavala, *Am. J. Sci.*, 301, 182 (2001). The current ratio is, of course, 1. If carbon dioxide in the atmosphere alone had controlled the temperature of Earth in the past, Earth must have been very much hotter than it appears to have been.

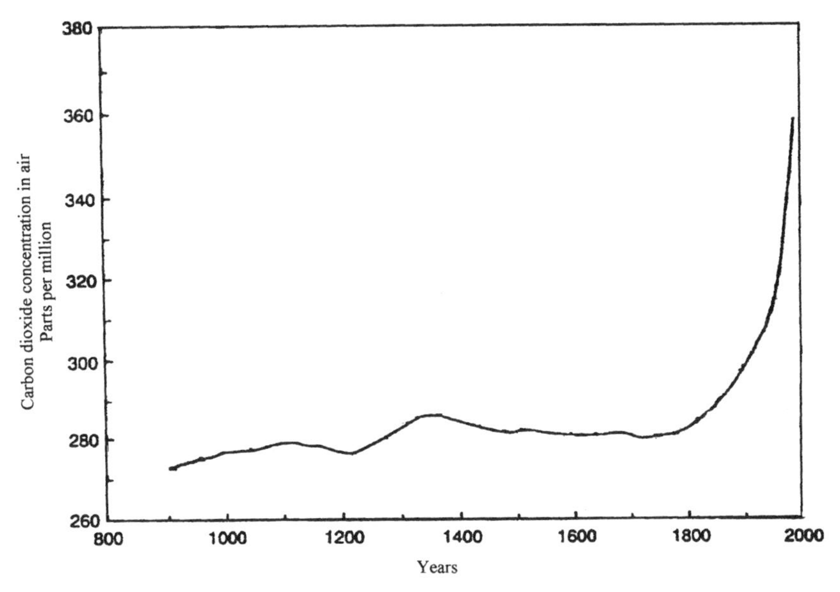

Figure 7.4. The increase of carbon dioxide in the atmosphere since the beginning of the Industrial Revolution. The curve represents an averaging of data from several sources, such as direct air measurements, ice cores, and sediments.

Nevertheless, the fact that one species of the ten million or so on this planet, *Homo sapiens*, in different ways uses about 45% of all plant growth must have a serious impact on the environment and all the ten-million-minus-one species attempting to exist alongside us.

What determines the ambient temperature today, including the phenomenon described by the term global warming, seems to depend in a complex fashion on a whole range of variables, including the levels of carbon dioxide, water vapour, and other gases in the atmosphere (the greenhouse effect), the luminosity of the sun, and the inclination of Earth's axis of rotation to the plane in which it turns about the sun. Only when all these are factored together can we be sure of what is happening and why. That said, carbon dioxide levels in the atmosphere are increasing and seem to have been doing so for about 150 years, or since large-scale industrialisation began (figure 7.4). The mean ambient temperature also seems to be increasing in some areas at least, though not everywhere. The precise correlation of temperature change and carbon dioxide levels is still a matter of dispute amongst authorities. Similar questions arise concerning the environmental implications of the presence of nitrogen compounds in the environment. Because ammonia is converted in the environment to nitrate and thence to nitrite and, at least in part, to nitrogen(I) oxide (N_2O) as well as to dinitrogen (see the global nitrogen cycle, figure 1.5), then all these compounds should enter into the discussion.

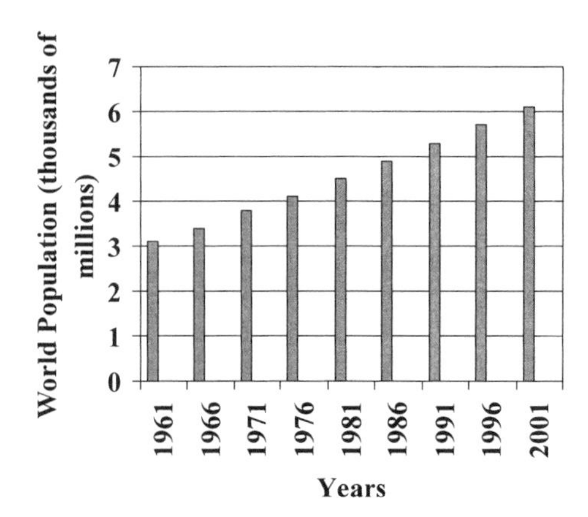

Figure 7.5. Growth in world population, 1961–2001 (based upon FAO estimates).

Population and Food Supply since the 1960s

If we look at what has happened to world population since the 1960s, the general predictions of the Club of Rome concerning growth appear to have been fulfilled. In 1960, the world population was about 3,000 million. By 1990, it had reached 5,000 million, and by 2000 the 6,000 million barrier had been crossed (figure 7.5). In early 2003, the world population had reached 6,316 million.[5]

All this was foreseen in the 1960s and is not far removed from the presumed exponential growth, but recently the rate of increase has been decreasing as time has passed. For whatever reason, average human fertility is dropping, though this average masks enormous differences. For example, in Western Europe, national populations are declining or are about to decline, whereas in Africa large increases are still evident. In 1989, the United Nations was expecting the population to reach over 8,000 million by the year 2025, but if the current trends are maintained, then the world population should stabilise at about 10,000 million by about 2050. There will certainly be problems in feeding this vast number of people, though whether they will be technical rather than political or economic is still a matter for discussion.

What is perhaps remarkable is that world food production is probably adequate to feed the current population, and there seems every reason to expect that the population of 2050 could also be adequately nourished. This has been brought about by the methods of intensive agriculture, with huge applications of fertilisers, pesticides, and insecticides, and especially of nitrogen fertilisers (figure 7.6). According to Smil,[6] perhaps 85 million tonnes of nitrogen were incorporated into harvested crops worldwide annually in the mid-1990s. This represents the sum of nitrogen from industrial and biological sources and is about as much nitrogen as was fixed annually by industry.

These numbers hide some rather interesting trends. Nitrogen fertiliser consumption actually reached a maximum in 1999/2000. The reason for the

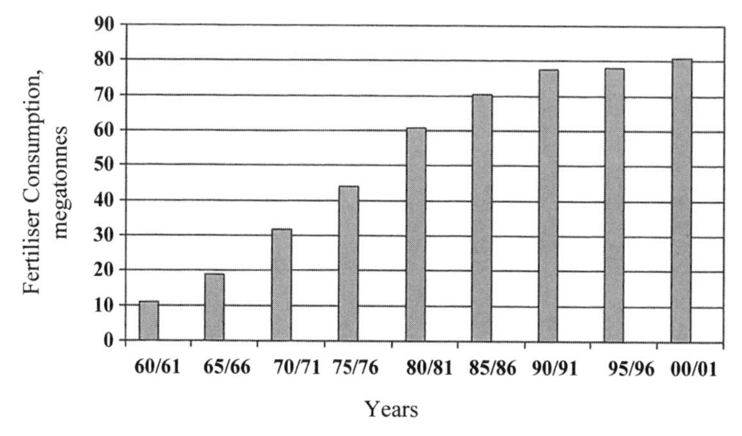

Figure 7.6. Increase in total world fertiliser consumption 1960/1961 to 2000/2001 (based upon International Fertilizer Industry Association data).

apparent decline since then is not obvious. Nevertheless, by 2004/2005 it has been forecast that consumption might reach 95 million tonnes, which will still leave a considerable excess of ammonia production capacity. Whereas in 1960/1961 the developed world consumed about 85% of the world nitrogen fertiliser (8.55 of the total 10.83 million tonnes), in 2000/2001 it consumed only about one-third (28.19 of the total 80.80 million tonnes). Total fertiliser (estimated in the customary way as nitrogen plus phosphorus oxide plus potassium oxide) consumption follows a parallel trend.[7]

Production of various staples such as wheat and other cereals in North America and Western Europe is far in excess of local needs, so much so that the areas under the plough are actually decreasing in several regions. That is not to say that the food being produced is not required elsewhere in the world. Just as with the distribution and consumption of rice in India, the problem is one of politics and economics, not of agricultural limitations. It has been estimated that about 40 million people die of hunger every year, although about 350 kg of cereal are produced annually for every human mouth requiring food (world cereal production in 2000 totalled just over 2,000 million tonnes, and the world population was close to 6,000 million).

Fertilisers and Food Production

How has this growth of food output arisen? According to The International Fertilizer Industry Association figures, the proportions of fertiliser actually used have changed. In the 1920s, about half the weight was phosphorus-based and only about a quarter nitrogen-based. This may have been due to the continuing influence of von Liebig, who was a great enthusiast for phosphorus fertilisers, and to the lack of a reliable industrial source of fertiliser nitrogen. By the 1990s, nitrogen fertilisers were about 55% of total world fertiliser consumption and phosphorus about 25% (figure 7.7). This has not stopped wars from being waged for sources of phosphorus fertilisers (phosphates), as in Western Sahara, but presumably the change to an emphasis on nitrogen is a consequence of the realisation that crop yields are more often nitrogen-limited rather than phosphorus-limited, that is, all the available nitrogen is consumed, and the plants then stop growing. What that implies is that the more nitrogen you add to the growing crop, the bigger will be the yield. The results of the increased application of nitrogen in world and Western European agriculture are shown in figures 7.8 and 7.9.[8]

More and more nitrogen is being applied to fields, though the chemical form of nitrogen used depends upon local commercial and social conditions. Sometimes, ammonium salts are used, sometimes urea is used, and sometimes, as in the United States, even liquid ammonia has been injected directly into the soil. World nitrogen fertiliser consumption was of the order of 80 million tonnes in 1999. Not unexpectedly, this consumption was not evenly spread throughout the world. Over 25% of it was produced and used in China, and as much as 40 million tonnes were produced and used in Asia as a whole. The major exporters of fixed nitrogen in that year were the United States and the members of the former Soviet bloc. Western Europe produced and consumed about 10 million tonnes.

Fertiliser production involves consumption of energy, often in the form of natural gas. The considerations above provoke the following question: Is the current rate of nitrogen fertiliser production (let alone any increased level) sustainable in the long term? In fact, fertiliser production currently uses less than 2% of the world's annual energy production and 5% of natural gas production. These quantities cannot be considered excessive if they result in an adequate food supply for the world population. Energy consumption in the manufacture of prepared foods is much more difficult to justify in these terms. According to the University of Wisconsin-Madison Center for Integrated Agricultural Systems,[9] in the United States only about 17.5% of the energy used to produce food ready for consumption is used "to the farm gate." Pro-

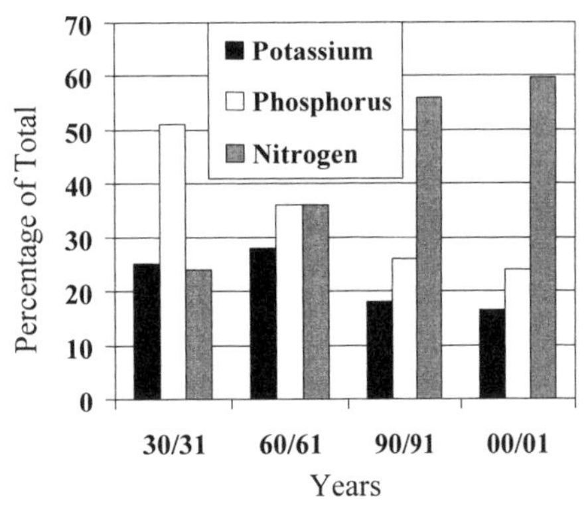

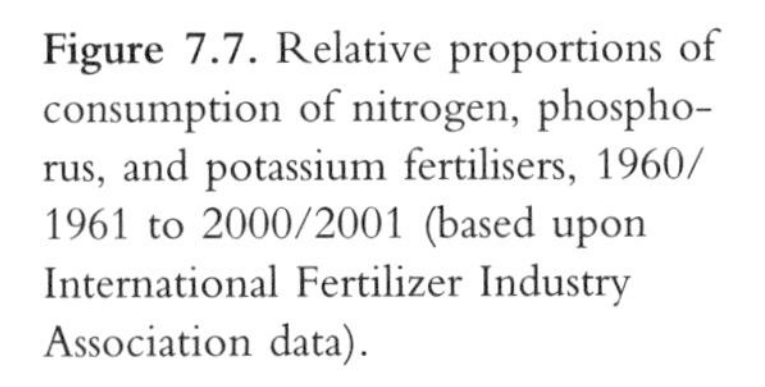
Figure 7.7. Relative proportions of consumption of nitrogen, phosphorus, and potassium fertilisers, 1960/1961 to 2000/2001 (based upon International Fertilizer Industry Association data).

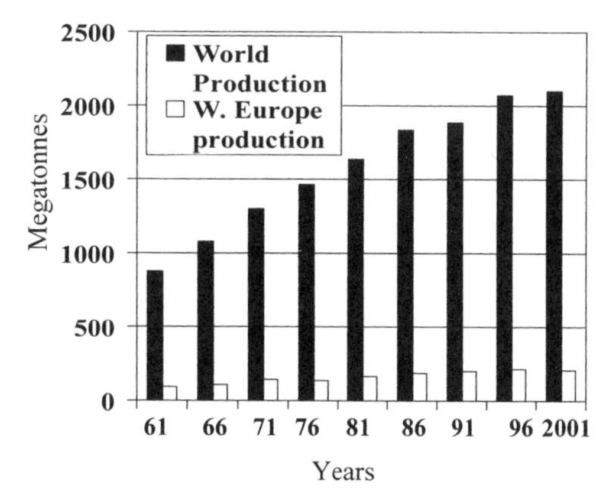

Figure 7.8. Production of cereals in the world and in Western Europe, 1961–2001 (based upon FAO data).

cessing industrially accounts for about 28%, distribution 9%, transport 11%, restaurants 16%, and home cooking 25%. These are rough estimates, but the United States uses three times as much energy per person for food than developing nations expend per person for all activities. About 10–15 times as much energy is used in producing food as is contained in the food itself. The figures for other developed countries may not be quite as skewed, but they aren't of a vastly different order. Food processing and cooking are not necessarily the benign processes we often assume them to be.

Chemical fertilisers account for about 30% of the energy used in farm food production in the United States. Since developing countries use so much

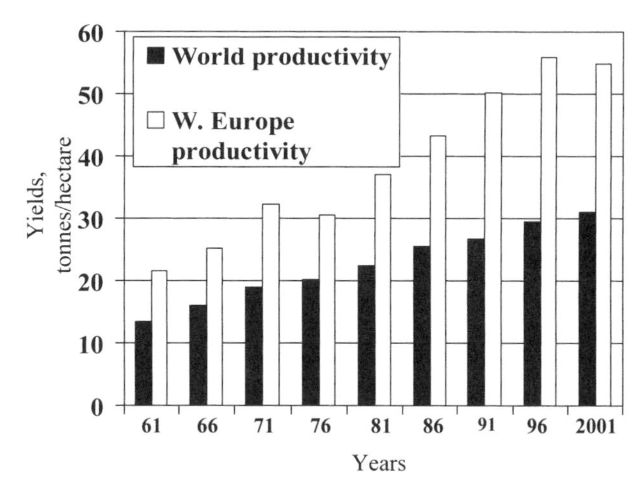

Figure 7.9. Average cereal yields per hectare in the world and Western European agriculture, 1961–2001 (based upon FAO data).

less energy than the developed countries, it is not surprising that global figures for nitrogen fertiliser production hide some astounding national variations. For example, in India the figure for natural gas consumption in fertiliser production is 40% of the local supply. However, it is calculated that the known world reserves of natural gas, if used solely for nitrogen fertiliser production, will last for at least another sixty years. Such considerations might lead one to conclude that everything must be, at least in the short term, well under control. If this is really the case, why are some people so agitated about nitrogen and energy?

There are several reasons. First, it is equally unwise to assume that a given pattern of consumption and production is likely not to change as it is to extrapolate consumption to the future exponentially. What is sure is that if present patterns of production and consumption are maintained, then the world will certainly need more fertiliser to feed the expected 10,000 million mouths that will inevitably be present by about 2050. "Organic" food production accounted for only about 2% of the total agricultural land use in the European Union (EU) in 2000/2001, and it is at least questionable by how much such methods could be expanded to feed the whole EU population. This implies a continued high demand for nitrogen fertilisers. Second, it is not at all obvious that the production and use of fertiliser itself is not without risk for the environment. Finally, what are likely to be the long-term consequences of distributing into the environment annually much more than the naturally fixed quantity of nitrogen? These are matters we consider next.

The Scale of Fertiliser Use

The application of nitrogen fertilisers to agricultural soils has generally increased steadily, at least until the 1990s. Quantities as large as 400 kg per hectare have not been unusual in intensive Western agricultural systems. Similar amounts of combined nitrogen contained in harvested material can be removed from fields, but generally lesser quantities are removed in crop harvesting than are supplied as fertiliser. According to Smil,[6] of the 85 million tonnes of nitrogen contained in the world crop harvest, 60 million are actually cropped, and 25 million remain in plant residues. A typical set of data for a crop in France suggests that 40–60% of applied fertiliser nitrogen is taken up in the above-ground part of a plant, whereas 20–50% will be left as plant residues, 5–20% will be left in the soil, perhaps as much as 30% will be denitrified, and perhaps as much as 10% will be lost by leaching.[10] Clearly, there is a wide range of possibilities. In summary, apart from harvesting, the balance of nitrogen, in the form of unused fertiliser and plant residues, remains in the soil to be transformed by natural processes and to be transported. Even most of the nitrogen removed in crops will reach the groundwater and surface waters via consumption of food, digestion, and subsequent excretion. Some of that will be denitrified from the soil and from sewage wastes. What are likely to be the effects of this increase of available nitrogen?

The quantities of extra nitrogen involved need to be seen in the context of the amount of nitrogen that might normally be available under natural circumstances. A soil will usually contain organic matter that consists, in part, of nitrogen. In a normal European cultivated soil, this has been estimated to be in the range 800–2400 kg per hectare. Undisturbed grassland has much higher reserves, and indeed when such grassland is freshly ploughed there is a large release of nitrogen into the groundwater, generally as nitrate. This is sometimes termed a nitrogen flush. What happens in detail depends upon the way in which the soil is managed and upon the underlying soil and rock structures and conditions. There is also an addition to the soil by atmospheric deposition as ammonia and oxides of nitrogen (NO_x, sometimes written as NOX). This is in the range 10–40 kg per hectare per year in Western Europe, and much of it is of natural origin. This accretion will always occur. Finally, the basic rate of biological nitrogen fixation needs to be considered. This again is a very variable quantity and globally amounts to perhaps 140 million tonnes per year. In particular circumstances, it can reach several hundred kilogrammes per hectare per year, but this will only be maximised with the appropriate legumes and microorganisms and in the absence of enough fixed nitrogen to cause the nitrogenase genes not to be switched on. What is evident is that any added synthetic nitrogen fertiliser will constitute only part of the nitrogen in the soil that is available to growing plants.

The era of intensive agriculture and the heavy application of fertiliser has seen an enormous increase in productivity, not only in the United States and in Western Europe but also in countries such as China. In most industrial economies, nitrogen fertiliser is considered to be cheap. In Western Europe and the United States, the price over the last twenty years has not varied very much from about $80 (U.S.) per tonne. However, the price is still an insuperable barrier to the provision of fertiliser to farming systems in poorer countries, where there is often no money to spare to purchase such apparent luxuries. In such circumstances, seeds are collected from crops at the end of one season for planting the next, and fertility is maintained as much as possible using traditional techniques.

So, what is the evidence that nitrogen fertilisers have really been of such tremendous value for feeding the world's population? The evidence is indirect but still very convincing. As shown above (figure 7.5), the world's population has increased dramatically since the 1960s. However, the increase has not been uniform worldwide. The populations of the United States and Western Europe have grown by about 48% and 18%, respectively, but the absolute numbers are relatively small. The corresponding figures for Africa, China, and India are 17%, 85%, and 120%, and there the aggregate is much greater.

Figure 7.10 shows the percentage trend in cereal production in various regions of the world from 1961 to 2001, taking the mean of the period 1989–1991 as 100%. For Africa, production has risen from about 46% in 1991 to 119% in 2001. All areas show a significant increase, but this is misleading. In

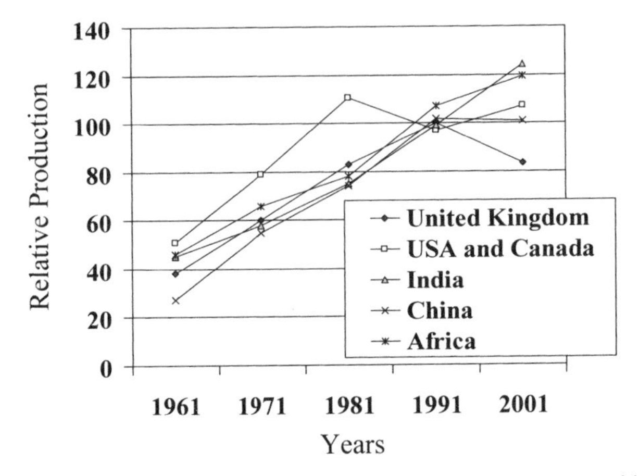

Figure 7.10. Trends in cereal production in different regions of the world, 1961–2001. These numbers are percentages, taking the mean of the years 1989, 1990, and 1991 as 100% in each case (based upon FAO data).

per capita terms rather than as gross percentages, food production and the cereal production in Africa are actually less in 2001 than they were in 1961! (figures 7.11 and 7.12) According to the Fertilizer Industry Association, the consumption of nitrogenous fertilisers rose from 0.41 million tonnes to just over 4.1 million tonnes in the same period. The corresponding United Nations Food and Agriculture Organisation (FAO) figures are rather larger but of the same order. There has been an increase in overall cereal and food production, but this has clearly been outstripped by population growth.

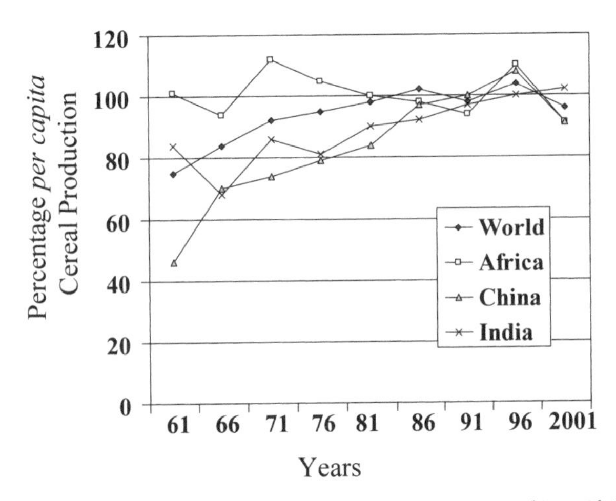

Figure 7.11. Per capita cereal production in the world and in Africa, China, and India, 1961–2001. These numbers are percentages, taking the mean of the years 1989, 1990, and 1991 as 100% in each case (based upon FAO data).

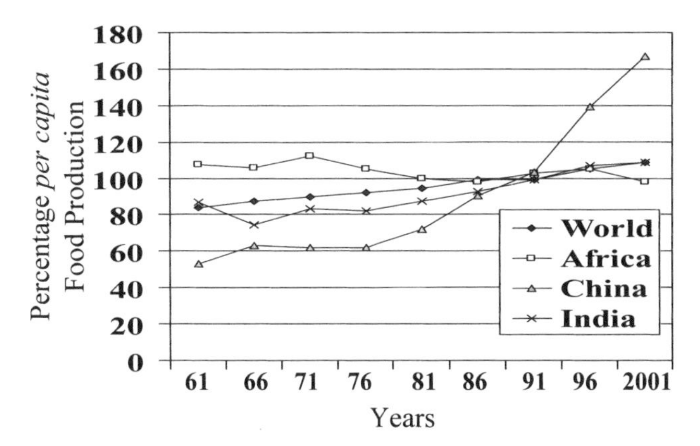

Figure 7.12. Per capita food production in the world and in Africa, China, and India, 1961–2001. These numbers are percentages, taking the mean of the years 1989, 1990, and 1991 as 100% in each case (based upon FAO data).

The contrast of Africa with both India and China is quite startling. Indian cereal and food production both increased from about 45% of the mean 1989–1991 value in 1961 to about 125% in 2000, but the per capita increase was much less, from about 85% to about 105%. Nitrogenous fertiliser use increased from about 0.25 million tonnes to about 12 million tonnes in the same period (see figure 7.13). The corresponding numbers for China are cereals 27–105% and food overall 31–177%. The per capita increases are less encouraging: for cereals 46–94% and for food overall 53–160%. During this time, nitrogenous fertiliser use increased from 0.54 million tonnes to an enormous 25 million tonnes (figure 7.13). Because a considerable proportion of grain production goes to feed animals, the effect of the increased nitrogen application is partly reflected in overall food production in the form of meat.

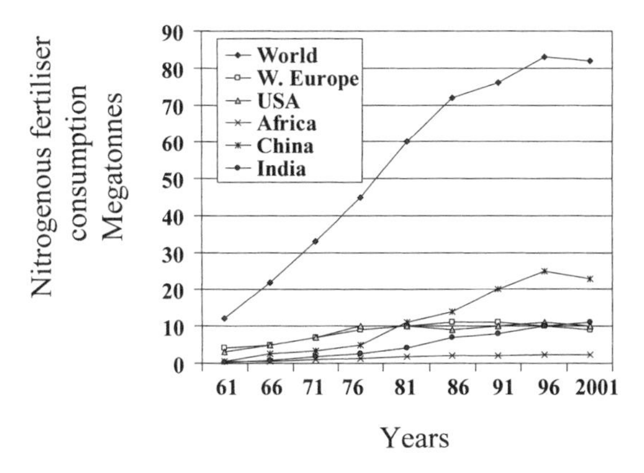

Figure 7.13. Consumption of nitrogenous fertilisers in various parts of the world, 1961–2001, in millions of tonnes (based upon FAO data).

Clearly, China has been much more successful in increasing its food production than India or Africa, but its use of nitrogen is correspondingly greater.

The increase in nitrogen fertiliser consumption in Canada and the United States, as well as in the United Kingdom, during this same period also has been impressive. In the former two countries, it has increased from about 3.1 million tonnes in 1961 to 11.7 million tonnes in 2001, though even more was used in the late 1990s. Both these countries are wheat-exporting countries. At the same time, cereal production increased from 51 to 107%, using the usual 1989–1991 mean as the basis. During the same period, the United Kingdom roughly tripled both its nitrogen fertiliser consumption and its cereal production.

What do all these numbers add up to? World food production has increased dramatically in the last forty years. So has fertiliser consumption, and especially nitrogen fertiliser consumption. In China particularly, each member of the population now receives on average about twice as much food as before, whereas in Africa the situation is as awful as it ever was, if not worse. India presents an intermediate situation. In North America and Western Europe, crop yields have also increased by a factor of more than two. These areas export cereals. Australia and Brazil have also increased their production of cereals dramatically between 1961 and 2001, and they now rival Canada in total production, though still only about one-sixth of that in the United States. North America remains the world's breadbasket and reserve larder. In general, experience in intensive cereal agriculture shows that there is a linear relationship between nitrogen fertiliser application and yield. Figure 7.14 summarises the changes in fertiliser application in the world as a whole, in developing countries, and in Western intensive agricultural systems. The drop in fertiliser application in the last group of countries since about 1980 is notable, but the yields per hectare in both categories of country have increased almost linearly with time. This is due, in part, to a more sophisticated application of fertiliser, which maximises uptake by plants and minimises washout in groundwater. It is evident that fertilisers based on

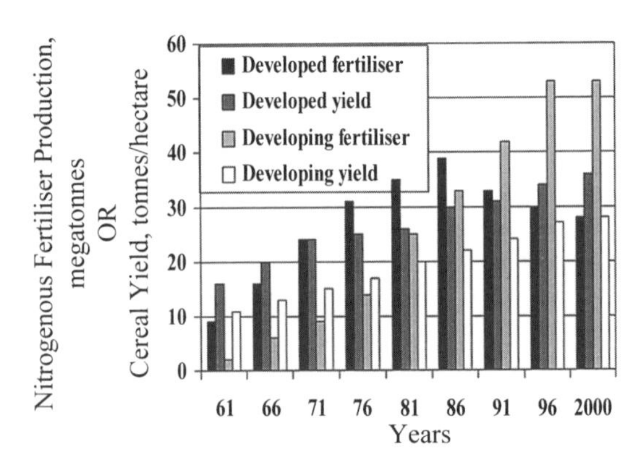

Figure 7.14. The relationship of cereal yields in tonnes per hectare to the total of nitrogenous fertiliser production in megatonnes for developed and developing countries at intervals during the period 1961–2000 (based upon FAO data).

nitrogen, so often the limiting nutrient, are used in much greater amounts than those based on phosphorus and potassium, corresponding to plants' lesser requirements for these elements.

Increased application of nitrogen fertiliser is not the only reason for productivity increases. To appreciate this fully, the criteria of yield and efficiency need to be defined very carefully. In terms of yields per unit area, the productivities of private plots and allotments are often much better than those of large, machine-worked fields. However, private plots and allotments do not contribute very much to the overall food balance in Western Europe or the United States. In contrast, in China they are very important. Such small plots consume very little energy apart from the energy expended in the manual labour of those who work them. In direct energy terms, they are very efficient. Against this, deep-sea fishing is probably the most energy-expensive way of obtaining food, especially if the fish are frozen and processed. Improved agricultural techniques and improved plant species (as employed in the Green Revolution) have undoubtedly led to increased yields per unit area in many agricultural systems. That some of these new plant types actually require fertilisers and pesticides to reach their maximum yields raises social, economic, and political questions that are not what this present discussion is about. The new varieties may well be of limited use for peasant farmers. What cannot be questioned is that artificial fertilisers, and especially nitrogen fertilisers, have favourably influenced crop yields.

Nevertheless, the impact of pesticides on food production has also been very significant. Despite the great reluctance of environmentalists to support the use of pesticides, their immediate ban without an alternative viable strategy to perform the tasks they currently perform could be disastrous. What is also very evident is that these materials are intended to be toxic, even if only to specific organisms, and their use should require the utmost care and evaluation. It may be difficult to disentangle the effects of the two kinds of additive, fertilisers and pesticides. The social, political, economic, and environmental consequences of restricting the uses of any of them need to be very carefully evaluated.

The longest-running scientific experiments anywhere in the world show this quite clearly.[11] Figure 7.15 shows the yields of various types of winter wheat in a series of experiments that have been running since 1843 without a break. These are the famous Broadbalk field experiments at the Rothamstead Experimental Station in England. These fields have been under continuous cultivation since the experiments started.

Figure 7.15 requires some explanation. It has not been possible to use the same variety of wheat for the whole period of the experiment, but the varieties used have been chosen to be as similar as possible, so that the plants' responses to the environment should not have changed significantly over the years. The yields of wheat in the completely unfertilised plots (lowest line through circles in figure 7.15), even without manuring, seem to have been remarkably constant for over 150 years at 1–2 tonnes per hectare. These yields are biologically

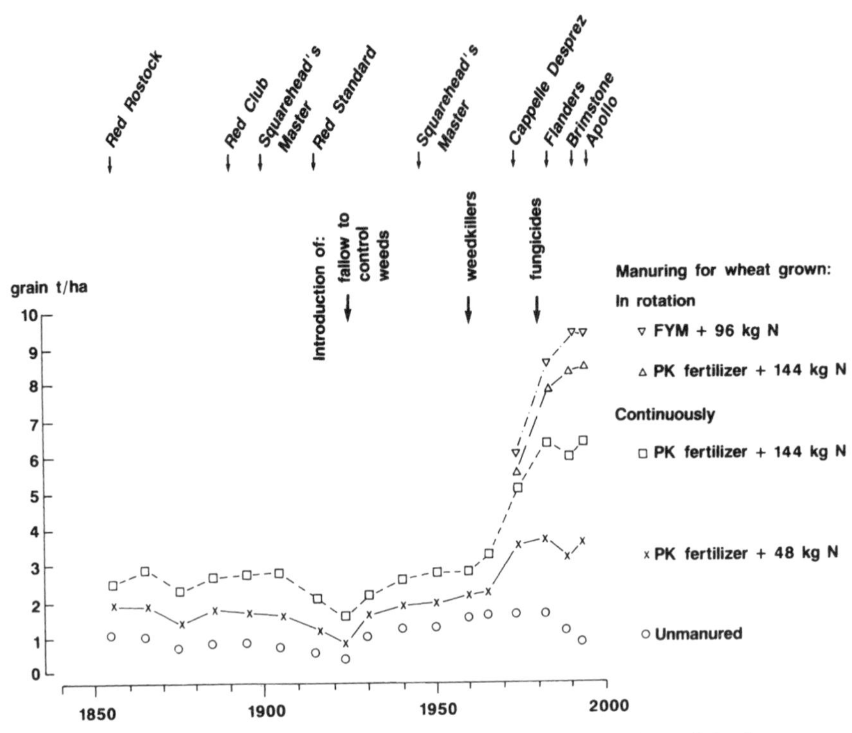

Figure 7.15. The Broadbalk experiment at Rothamstead Experimental Station. This experiment has lasted more than 150 years, and the parameters have changed from time to time. The variety of wheat employed has had to be changed, and attempts were made at different times to control weeds and pests. These are marked on the figure. Originally, only continuous cultivation was employed, but since about 1970, rotations have been introduced into the experiment. The effect of introducing weedkillers seems particularly dramatic. Reproduced by courtesy of Professor A. E. Johnston from Proceedings No. 459, International Fertilizer Society, York, U.K., 2001, p. 34. (FYM is farmyard manure; PK fertilizer contains principally phosphorus and potassium.)

sustainable but not economically viable in an intensive agricultural economy. What is evident is that fertilisers increased the yields by a factor of about two or more (lines through crosses and squares in figure 7.15). Crop rotation is also beneficial, but whether that is useful in a market that requires constant and assured supply is another matter. The more modern plant varieties seem to require fertilisers rather than merely benefit from them, and yields drop if fertilisers are not applied, but they are evidently much more responsive to added nitrogen than the older cultivars. The drop in the unfertilised yields in the 1920s was due to buildup in competition by weeds, and this was controlled then by leaving the plots fallow for one year in five, an echo of a practise literally centuries old. Of course, in economic agriculture rather than in experiment, fertilisers are not the only factors that affect yields, there being greater or lesser perturbations due to the application of weed killers and pesticides, changes in the depth of tillage, water supply and temperature variations, and other alterations. Nevertheless, the accumulated evidence shows quite clearly that a major influence in increasing crop yields is the provision of fixed nitrogen. For example, according to Smil,[6] Chinese double-cropped paddies in Hunan yielded less than 2.5 tonnes per hectare of rice in 1950 with no fertiliser. Now they receive as much as 400 kg of nitrogen per hectare and yield 6 tonnes per hectare in a single crop. It is estimated that 40% of all people alive today are dependent upon Haber–Bosch nitrogen, and without it probably less than half the current world population would be adequately fed.

Of course, it is inadvisable to restrict the discussion to a single kind of agricultural product, cereals. That is why figures for overall food production have been quoted. Within that broad category, there are some interesting variations that do not, however, change the total picture. According to FAO figures, soybean production in Africa increased from about 72,000 tonnes in 1962 to 1.1 million tonnes in 2001. The yields per hectare tripled. For developing countries as a whole, the figures are 7.4 and 93.5 million tonnes, with a yield more than triple, whereas for developed countries the corresponding figures are 19.5 and 83 million tonnes, with a yield rather less than double. This confirms that developing countries can improve yields significantly and can sometimes outperform developed ones, but one also needs to ask what happens to this increased output. It is not always to the general benefit of the indigenous populations.

What Is Wrong with Nitrogen Fertilisers?

If this is self-evidently true, why do people regard nitrogen in the environment with such suspicion? Why do environmentalists warn of the dangers of nitrate and pesticides and encourage us to adopt "organic" methods of producing food? The problem of pesticides cannot be discussed here, but to deal with these questions as far as they concern nitrate, we need to know what happens to nitrogen in plants and in fertilisers under normal field conditions.

Figure 7.16 shows what happens to nitrogen when it is applied in different situations in a temperate climate. Some is absorbed by the plants and is harvested and removed from the field, and this may then have value to humans or animals as nutrient. Clearly, the proportion removed in this way should be maximised. Some of this nitrogen may return in another form as manure, of human or animal origin, but that is not important for this argument. Some is washed out in the groundwater with the rains. Other possible fates include runoff in surface water, denitrification, and immobilisation on soil organic matter. Figure 7.16 shows that grassland loses practically no nitrogen by leaching. Yet more nitrogen remains in the unharvested plant residues in the fields. As these decay, the nitrogen compounds are converted by bacteria to nitrates, and they also wash out into the groundwater. Bare fallow loses all the added nitrogen as well as soluble nitrogen produced by decay. This is because very few plants are around to take it up. Cultivated areas take up

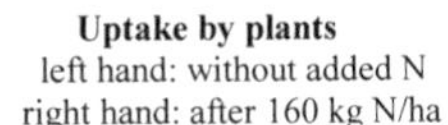

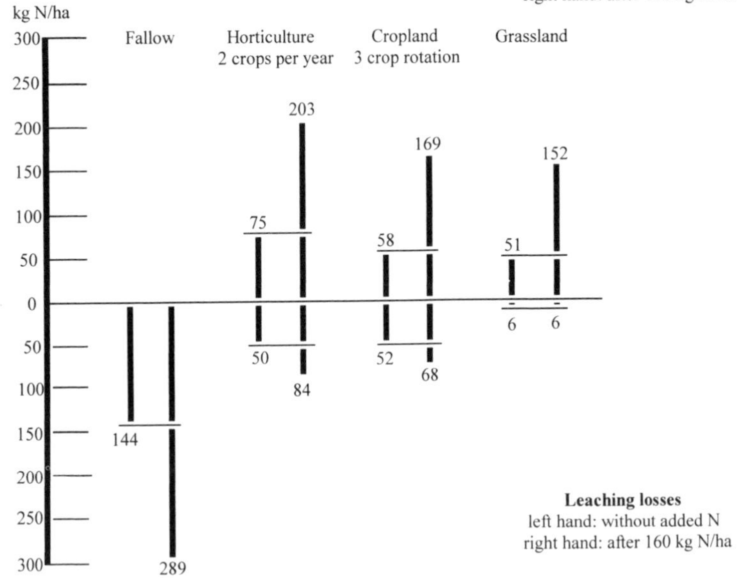

Figure 7.16. The uptake of applied nitrogen by various crops, adapted from various sources but principally from data taken from Dressel, Jung, and Will. These authors worked on various aspects of plant nitrogen uptake, leaching losses of available nitrogen, and denitrification. As an example, see J. Dressel, J. Jung, and H. Will, *Gartenbauwissenschaften*, 49, 106 (1984). For a more recent discussion, see V. S. Prakasa Rao and K. Puttana, *Curr. Sci.*, 79, 1163 (2000).

some nitrogen but also lose some by leaching, intermediate between grassland and fallow. It is worth noting that when grassland is first ploughed, there is an enormous loss of nitrogen by leaching. These processes produce a characteristic nitrate "flush" that, in temperate climate areas, is at a maximum in the winter and early spring. This is when the rainfall is at a maximum and when there are few plants growing to take advantage of the availability of the nitrate (figure 7.17).

Though the seasonal variations are clearly evident, the baseline shows a significant though gradual upward trend until about 1990. Since that time, the baseline in many rivers shows an improvement in nitrate levels, but the rate and frequency of fertiliser application in the United Kingdom has changed considerably since then. Figure 7.18 shows that the nitrate levels in U.K. rivers have at least been stabilised. Nevertheless, a considerable proportion of nitrate in most U.K. rivers is still of agricultural origin.

The baseline variations can usually be explained by the changes in the rates and manner of fertiliser application undertaken to reduce nitrate runoff, as well as by better wastewater treatment. However, in a chemical sense there is no such

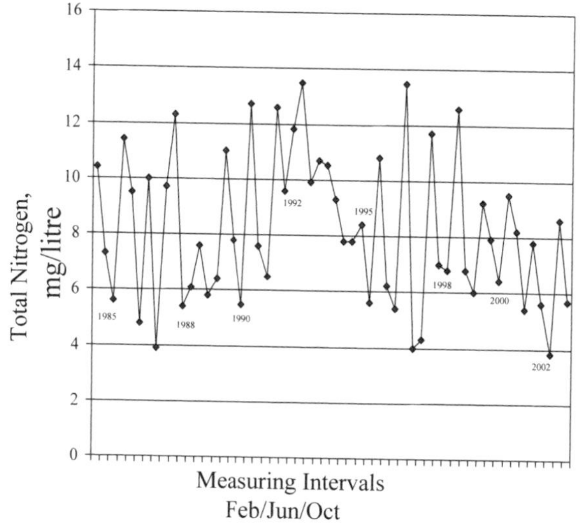

Figure 7.17. The variation in oxidised nitrogen (effectively nitrate) concentrations from a sampling point in an English river, the Great Ouse. (Nitrate levels may be calculated by multiplying individual nitrogen values by 4.43.) The values plotted were those determined in February, June, and October each year. Neglecting occasional perturbations perhaps due to local effects, the diagram shows the seasonal changes in nitrate levels, generally at a maximum in February. Until about 1995, there appears to be a rising trend of overall nitrate levels, continuing the tendency recorded since at least 1960. The diagram suggests that since about 2000 the trend may have been reversed. On several occasions, the nitrate level rose above the WHO recommended maximum, 11.3 mg/L nitrate nitrogen. The data for this figure were kindly supplied by the U.K. Environment Agency, Peterborough.

thing as "organic" nitrate. As Gertrude Stein certainly didn't say, though she might have done so had she any training in chemistry: "Nitrate is nitrate is nitrate." If one were to replace all the synthetic nitrogen fertiliser applied to a field with enough ("organic") manure to provide a similar quantity of nitrogen (and one would need to do this to maintain the crop yields), then the amount of nitrate washed out into the groundwater would probably be very similar. It might not be released at the same rate as nitrate from "artificial" fertiliser because it would need to be liberated by a process of mineralisation from the dung or organic matter in which it was originally combined. The rate of release would depend upon factors such as the climate, temperature, and rainfall. In any case, even "organic" produce in industrialised countries obtains a considerable proportion of its nitrogen via aerial deposition of nitrate generated non-organically. This has been calculated to be in the range 20–40 kg per hectare in a typical European situation, where fertiliser application of nitrogen might be in the range 100–250 kg per hectare.[12]

The grounds for trying to reduce nitrate release are as follows. First, it is claimed that continuous application of synthetic fertilisers causes soil

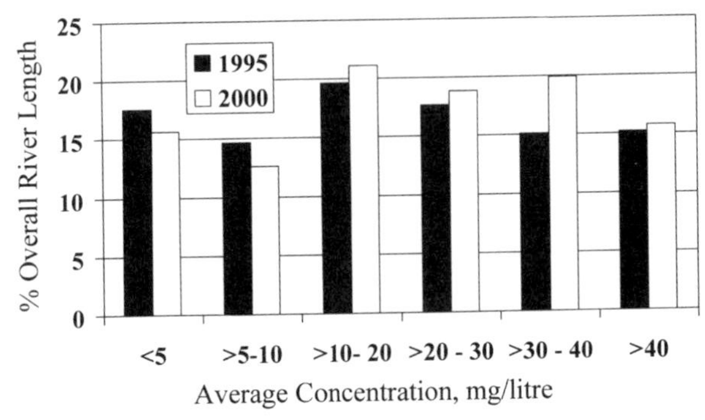

Figure 7.18. The trend in nitrate levels in selected U.K. rivers. The data represent the summary of average nitrate concentrations determined in many rivers, covering a total length surveyed of about 40,000 km. Since these are average figures, they do not show the greatest levels recorded, but these must sometimes have exceeded the WHO recommended level of 50 mg/L NO_3^-. Despite efforts to control nitrate levels, there seems to have been little significant change between 1995 and 2000, though 2000 was a particularly wet year. The data for this diagram were taken from The Environment Agency, *Rivers and Estuaries—A Decade of Improvement*, Environment Agency, Bristol, 2002.

degradation and a loss of wildlife and wild plants from the general environment. This would be difficult to deny plausibly. The organic material provided by manures also generates humus that helps maintain the quality of the soil. Such an external source of organic matter is absent if only synthetic fertilisers are used. Second, it is claimed that the nitrate runoff causes eutrophication (nutrient enrichment) in groundwaters (i.e., water held within the soil and rocks below the water table) and surface waters (i.e., water held in rivers, ponds, lakes, and reservoirs), and this leads to rapid algal growth, dioxygen depletion in the water as the plant material decays, and all the unpleasant consequences for plants, fish, and other aquatic animals. Third, it is suggested that high nitrate concentrations in drinking water need to be reduced to avoid consequent increases in the incidence of stomach and gullet cancer. Fourth, it is said that such high nitrate concentrations can also lead to "blue baby syndrome" in very young babies (neonates). Finally, the natural process of denitrification inevitably leads to the generation of some nitrogen(I) oxide, N_2O, and this damages the ozone layer and enhances global warming. However, all these claims need to be treated with some caution.

Soil Degradation and Environmental Damage

The changes wrought upon the environment by agricultural practices are clear for all to see, and not only in the industrialised countries. However, it is important to realise that humans have been imposing change upon the landscape ever since they decided that agriculture is a better way of providing food than hunting and gathering. The growth of deserts and deforestation may be just as much a consequence of human activities as the dust storms arising as a result of ploughing prairie or steppe. Nevertheless, in order to understand what is happening now, we have to try to uncouple the effects of our activities from those changes wrought by Nature. It is incorrect to imagine that at some time before the Fall of man the world was ever in some kind of equilibrium. Change is constant. The fabled English landscape with its hedges and small fields, so beloved of poets, painters, and manufacturers of chocolate boxes, is a product of human activities. Global warming may also be such a product. However, our very presence here on Earth in such vast numbers means that we must influence the environment in one way or another. People who build energy-efficient "green" houses may be protecting the environmental status quo to some degree, but inevitably the new house will affect the environment. The question is not whether the environment is affected, because it certainly will be, but by how much or how little. This is an inevitable consequence of our existence on Earth.

With that reservation, it is evident that the development of vast areas of monocultures as in North America must have reduced the possibilities of suitable environmental niches for many wild plants, animals, insects, and bacteria to occupy. How you evaluate these changes depends very much on whether you are a farmer, a consumer, or an environmentalist. If you are more than one of these, and this is true of most of us, then the evaluation is far from straightforward. The changes in agricultural practice have been great but quite subtle. In the United Kingdom, there has been a change from mixed arable/dairy farming in the areas with the best soils to monocultures, be they of barley, wheat, rape, or even linseed. Even within the monocultures there have been changes from one crop to another, many a direct result of political and economic pressures rather than a rational response to circumstances. In some parts of the world, rotation of monocultures may be practised, but this may not be the case with wheat, corn, barley, and rice. The conclusion that agriculture causes a loss of habitat and consequently of varieties of all kinds of living things seems inescapable. Even though the direst effects can be alleviated by various strategies such as leaving uncultivated margins around the edges of fields, there will always be a major perturbation of the environment.

The related problem is that continual use of the same soil for raising the same crop year after year is said to lead inevitably to soil degradation. The problem is to define exactly what one means by soil degradation. One important factor is that continual cropping and removal of plant residues (for example, by burning) would be expected to cause a drop in the organic material content of the soil. This undoubtedly happens, even if burning can also kill pests and recycle some nutrients more rapidly than happens as a result of decay of plant material. The problem can be alleviated only in part by improving techniques of soil management. One result of intensive agriculture might well be the dust-bowl conditions that have been experienced both in the United States and in the former Soviet Union. Another long-term consequence may be the loss of trace elements as the crop extracts its requirements over the years. In addition, the continued application of high levels of fertilisers can lead to the buildup of concentrations of certain elements, such as phosphorus, potassium, and nitrogen, in the soil. This will also affect soil structure and the populations of species such as bacteria. How damaging all the changes might be is open to question. Some soils are not terribly productive and fertile in any case. However, very little of what we have discussed here can be ascribed directly to nitrogen rather than to the effects of farming. It would be unwise to refrain from using nitrogen fertilisers on the basis of these considerations alone. Indeed, some of the consequences can be alleviated by applying fertiliser only at times when plants are growing, and then only in quantities to satisfy the plants' immediate requirements.

Nitrates, Nitrogen, and Eutrophication

Further problems may arise once nitrogen compounds pass into groundwaters. Whatever the form in which fertilisers are applied—as nitrate directly, as salts of ammonia, as ammonia itself, as urea, or even as manure—the action of the air and of soil bacteria is likely finally to furnish ultimately a dilute solution of nitrate. Nitrates in general are normally very soluble in water, and so will tend to wash out of the soil in the rain or with irrigation. The consequence is an increased concentration of nitrogen, not only in groundwaters but also in surface waters and eventually in the sea. The question this poses is "Does this matter?" and the answer, as always, is "It all depends."

A certain increase in nutrient levels in waters may be advantageous, for example, an increase in plant growth might lead to an increase in the supportable fish population and hence to a larger supply of fish as food for humans. However, undesirable (in human terms) fish and animals might also take advantage of the increased food supply. In addition, if the increase in nitrogen and other nutrients becomes excessive, then other troubles may ensue. What actually happens is generally determined by the nitrogen to phosphorus (N/P) ratio, and it should be borne in mind that this is affected not only by fertiliser washout but also by additions from domestic sources such as sewage and detergents. However, the determinations of nitrate in river flows in temperate regions may well lead to an underestimate of the real nitrogen flux. Recent evidence of the nitrogen flux from some unpolluted temperate forests shows that most of the soluble nitrogen leaves in the form of organic nitrogen compounds, not as nitrate or ammonium ions.[13] Organic nitrogen compounds may be more generally involved in nitrogen fluxes in temperate regions than hitherto believed.

These outflows from many environments certainly contain considerable quantities of phosphorus as well as nitrogen. Most of these sources of phosphorus are the result of human activities. Phosphate itself is not very mobile and tends to stay adsorbed to soil and rocks. What limits the growth of plants, particularly of algae, in waters is normally the availability of phosphorus, not nitrogen. In temperate lakes, the ratio of nitrogen to phosphorus is normally of the order of 20 to 1. Only when the nitrogen to phosphorus ratio in freshwaters is less than about 4 to 1 is nitrogen limiting, so the best way to avoid this situation is to remove phosphorus, an easier option than removing nitrogen. However, if phosphorus availability is increased as a result of human activity, such as industrial production or washing clothes with detergents, and there is also a good supply of nitrogen, then rapid growth of plants (algae) may occur in fresh as well as in marine waters. However, plant growth is affected by many factors in addition to the presence or absence of nitrogen and phosphorus. Amongst these are the temperature and the concentration of iron. It is therefore unwise to make inferences about causes simply from the observation of rapid plant (algal) growth.

Although moderate increases in algal growth may have advantageous effects, such as an increase in fish population, excessive growth may have unpleasant effects. The algae may shield other growing things from light. When the algae die, they will inevitably decay, and this consumes dioxygen otherwise dissolved in the water. The decay can cause unpleasant smells and produce toxins, and the lack of dioxygen may kill both animals and plants. This is the process popularly known as eutrophication, and, at least in freshwater, a predominant species involved is comprised of nitrogen-fixing cyanobacteria.

Eutrophication in the Seas

The problem of eutrophication extends to all types of water reservoirs, including the sea. According to a group of experts[14] from a wide range of organisations assembled to evaluate aspects of marine pollution: "The rate of introduction of nutrients, chiefly nitrates but sometimes also phosphates, is increasing and areas of eutrophication are expanding . . . Two major sources of nutrients to coastal waters are sewage disposal and agricultural run-off." As noted earlier, the annual rate of industrial fixation is of the same order of magnitude as natural fixation. Eutrophication and algal blooms have been seen off the coast of Peru, in the Baltic Sea, around the coasts of the United States, and in the Gulf of Mexico. In the last case, this seems to be seasonal and is caused by a combination of factors, including the flush of nutrients brought down by the River Mississipi, the stratification of the freshwater and the seawater, and solar warming in the summer.[14]

However, the rate of application of fertiliser is declining in the United States as it is in Western Europe, and it is far from certain that the blooms often seen in the Pacific (and associated with the El Niño phenomenon) are always provoked directly by artificial fertilisers running off the land.[15] The major polluting runoff from the United States probably is of agricultural origin, but there are large variations from place to place. Sewage is not an important source except very locally. The inference from data such as those summarised here is that marine eutrophication is sometimes a natural phenomenon and that attempts to reduce nitrogen runoff need to be carefully assessed. The health of the sardines off Peru may have profound consequences for sardine fishermen and for organisms in the sea but does not necessarily affect the health of the population as a whole. A proper cost–benefit analysis needs to be carried out for each proposal designed to remedy eutrophication, with due application of the precautionary principle: if in doubt, don't do it. However, a good cost–benefit analysis requires good information, and this is not always available. It should be borne in mind that agriculture is not necessarily the major source of polluting nitrate and that the drop in productivity and the increase in food prices that could occur upon an unconsidered major reduction in fertiliser application needs to be factored into the equation.

Pollution in Rivers

Rivers are the immediate recipients of much of the runoff from agricultural land. Rivers certainly have suffered from pollution by human activities, and this has been demonstrated many times by the low levels of observed dioxygen concentration and the effect on plant and animal life. However, in some developed countries, things are not as bad as they used to be. Figure 7.17 shows the seasonal flux of nitrate in one U.K. river up to 2002. The amount of nitrate involved is a product of both the concentration of nitrate and the water flow, which is greater in winter and spring.

As discussed above, the decay of plant material at a time of year when few plants are growing results in a nitrogen flush. Whether this is damaging or not requires some consideration. In addition, since 1990, less nitrate fertiliser has been applied to fields in the United States and in Western Europe, and the observed levels of nitrate in some rivers have dropped (see figure 7.17). However, the nitrogen flush is a natural phenomenon that will occur even in the complete absence of agriculture. The problem is again one of the degree to which a nitrogen flush can be tolerated rather than whether the phenomenon can be suppressed entirely. If there is no significant plant growth in a river, as is probable in winter, the high levels of nitrate might well be flushed away to the sea without having any great influence on dioxygen levels in the river, though the same would not be true of enclosed stagnant waters, such as lakes.

Certainly, by modifying agricultural methods it seems possible to effect improvements. However, not all observations should be taken at face value. One technique is to apply fertiliser in small amounts but at regular intervals, particularly when the crops are actually growing and taking up nitrate. For example, in a well-publicised experiment, this kind of approach was used in the area of the chalk hills called the South Downs in the south of England. Much of the drinking water used in this region comes from artesian wells because the chalk acts as an enormous sponge and can store vast quantities of water. The levels of nitrate in the water taken from some of these wells were adjudged to be dangerous on health grounds.

The modified agricultural techniques had an immediate effect on the nitrate levels in the drinking water taken from the artesian wells, and that raises an important question: Water takes a long time to percolate through the chalk of the Downs and to emerge in the streams below the chalk beds, so why was the response to the new methods so rapid? It could be that faults and cracks in the chalk allowed the polluted water to be drawn very quickly. It could also be that what was observed resulted from changes made twenty years earlier. Questions such as these require more, and more intensive, study.

Nevertheless, it is undoubtedly true that, when the rivers are used as sewers, the quality of river water always suffers. Pollution from heavy metals and from bacteria can be very worrying and dangerous, but there seems to be little evidence of nitrate being a prime cause of eutrophication in rivers, at least in temperate regions. It may be so in closed systems such as reservoirs and lakes, but the effect is not common in most developed countries and should be in any case relatively easy to control, particularly if high phosphate levels are also present. Our current interest is not phosphate but nitrate, and the direct ill effects of nitrate seem to be negligible. It is not poisonous to plants or animals.

Nitrate and Health

The maximum "safe" level of nitrate in drinking water is accepted in most countries to be 50 milligrams of nitrate per litre, 50 mg/L. The WHO recommends this level, and it is applied throughout the European Union. An alternative way of expressing this is not as nitrate but as nitrate nitrogen, for which the safe level then becomes 11 mg/L. These two limits are effectively the same (try calculating the nitrogen content of nitrate, NO_3^-), but it is necessary to be clear what number is being used in any discussion, that of nitrate itself or nitrate nitrogen.

Two principal health dangers are generally associated with nitrate.[15,16,17,18] One is called "blue baby syndrome," or, more scientifically, methaemoglobinaemia. The other is stomach cancer. The former condition is believed to arise as follows. Although humans have well-developed systems in their bodies for controlling acidity, such systems do not operate well in very young and newborn babies. The result can be that the body of a baby that receives a high dose of nitrate when the baby is, for some reason, more alkaline than it might normally be. This in itself is not a dangerous condition. However, under alkaline conditions, bacteria occurring naturally in the body can reduce nitrate, NO_3^-, to nitrite, NO_2^-.

The agent that carries dioxygen around the body in the blood is, of course, haemoglobin. This it is able to do without becoming oxidised. It is simply a carrier, and binding the dioxygen to it causes no chemical change in its constitution. This is important because the oxidised form of haemoglobin, called methaemoglobin, cannot carry dioxygen. If an excessive amount of the haemoglobin becomes oxidised, and if, as with a newborn baby, the body cannot easily reduce the methaemoglobin back to haemoglobin, the blood cannot carry dioxygen to the muscles and wherever it is required, and the body begins to suffer from lack of oxygen. The problem is that nitrite is a good catalyst for the oxidation of haemoglobin by dioxygen, and in unpropitious circumstances the body begins to suffer. The consequence for a baby can be fatal. The child literally suffocates, turning blue in the process (cyanosis). Hence the name of the condition, blue baby syndrome. However, another name, which arises from the original definition of the condition and is equally informative, is well-water methaemoglobinaemia.

What is the incidence of this condition?[19] Apparently, it was first recognised in the United States in the 1940s when it was associated with a source of water from a contaminated well. The last confirmed death from the condition in the United Kingdom was in 1950. The incidence in Eastern Europe is considerably higher but generally less than 1 in 100,000 and totalling perhaps several hundred cases in the last fifty years. What is not evident is whether the condition would have occurred if the water sources had not been contaminated with bacteria, which was true in many of the cases reported. Wells in proximity to farmyards may indeed contain water with high levels of nitrate. They may also contain high levels of pathogenic bacteria, and either nitrate or bacteria might be the cause of the methaemoglobinaemia.

Of course, death may be only the most dramatic result of an attack of methaemoglobinaemia. There have been several reports of illness arising from eating vegetables such as carrots and spinach, and vegetables often contain considerable quantities of nitrate. Carrot soup was until recently considered a suitable treatment for babies suffering from diarrhoea. However, in most cases the bacterial infection arising from lack of hygiene in the preparation of the vegetables may have been to blame rather than nitrate. The same applies to the illnesses that can arise when a baby is fed with milk formula rather than with the mother's milk. Unless strict hygiene is observed in the preparation of the formula, it can make the baby unwell, but it is not by any means sure that nitrate in the water would be to blame.

However, if we assume that nitrate alone was responsible for all the reported cases of methaemoglobinaemia apparently caused by drinking water, what is the maximum safe level of nitrate? In the fatal case of 1950 in the U.K., the level was reported to be 200 mg/L. In another non-fatal case in the same year, the level was 95 mg/L, but the water was apparently also contaminated with coliform bacteria. In both instances, the water source was a well. In fact, it is doubtful whether any case has arisen from water coming from a municipal water supply.

When water is bacterially contaminated, cyanosis has been observed at nitrate levels as low as 50 mg/L, and this is essentially why this level of nitrate has been set as the highest permissible level in drinking water. However, for a clean water supply, there seems no reason why such a level should be set. Because it is very expensive to remove the excess of nitrate from drinking water, it is at least arguable that the money might be better spent in ensuring that everyone has a clean supply of water not contaminated by bacteria. There is no hard evidence that high levels of nitrate alone are, in themselves, of any danger to health. Large quantities can be consumed without apparent ill effects. Bearing in mind that perhaps 90% of the normal domestic supply is used for cleaning clothes or for flushing toilets and is never imbibed, consideration of other strategies to eliminate the few cases of blue baby syndrome is, at the least, sensible.

The other major health hazard of nitrate is said to be stomach cancer. The danger arises, once again, from nitrite rather than nitrate. If nitrate is reduced bacterially to nitrite in the mouth and then mixes with food in the stomach, then the reaction with amines in the food can produce substances called nitrosamines. These are known to be highly carcinogenic. Nitrite itself has the ability to make raw meat look attractively red and has been used in the past in smoked meats and sausages for this purpose. It is no longer considered to be a sensible additive.

Although nitrate levels in groundwaters and in some drinking water supplies have increased, death rates from stomach cancer in the United States have decreased from about 40 per 100,000 in 1930 to about 5 per 100,000 in the 1990s for men. Women seem to be less susceptible to stomach cancer, but even so their death rate seems to have dropped from about 30 per 100,000 to about 3 per 100,000 in the same period (figures 7.19a and b). In absolute terms, there are cancers to worry about that are much more damaging than stomach cancer! Figure 7.20 shows this very clearly. However, as a caveat, it should be noted that death rates may not parallel rates of incidence. For example, rates of incidence may increase due to better diagnosis, and death rates may fall due to improved treatment. Nevertheless, the death rates are a good first indicator of the importance of a given variety of cancer.

In Denmark, where the consumption of nitrite-doped smoked meats was high, the incidence of deaths from stomach cancer dropped from about 70 per 100,000 population in the 1930s to about 20 by the mid-1980s. The Organisation for Economic Cooperation and Development recorded a death rate for stomach cancer in Denmark in 1995 of 5.5. The World Health Organisation concluded in 1985[20] that there "is no convincing evidence of a relationship between gastric cancer and the consumption of drinking water containing nitrate levels up to 10 mg nitrogen per litre (essentially 50 mg/L nitrate). The U.S. National Research Council concluded[21] that "exposure to the nitrate concentrations found in drinking water in the United States is unlikely to contribute to human cancer risk."

A British study looked at the amount of nitrate in the bodies of individuals from areas known to be of high risk of suffering from stomach cancer and others from areas of low cancer risk. In fact, those persons with the highest concentration of nitrate in their saliva seemed to come mainly from the areas of low risk of stomach cancer and vice versa.[22] At least one other British study of mortality of workers in a nitrate fertiliser factory is consistent with this

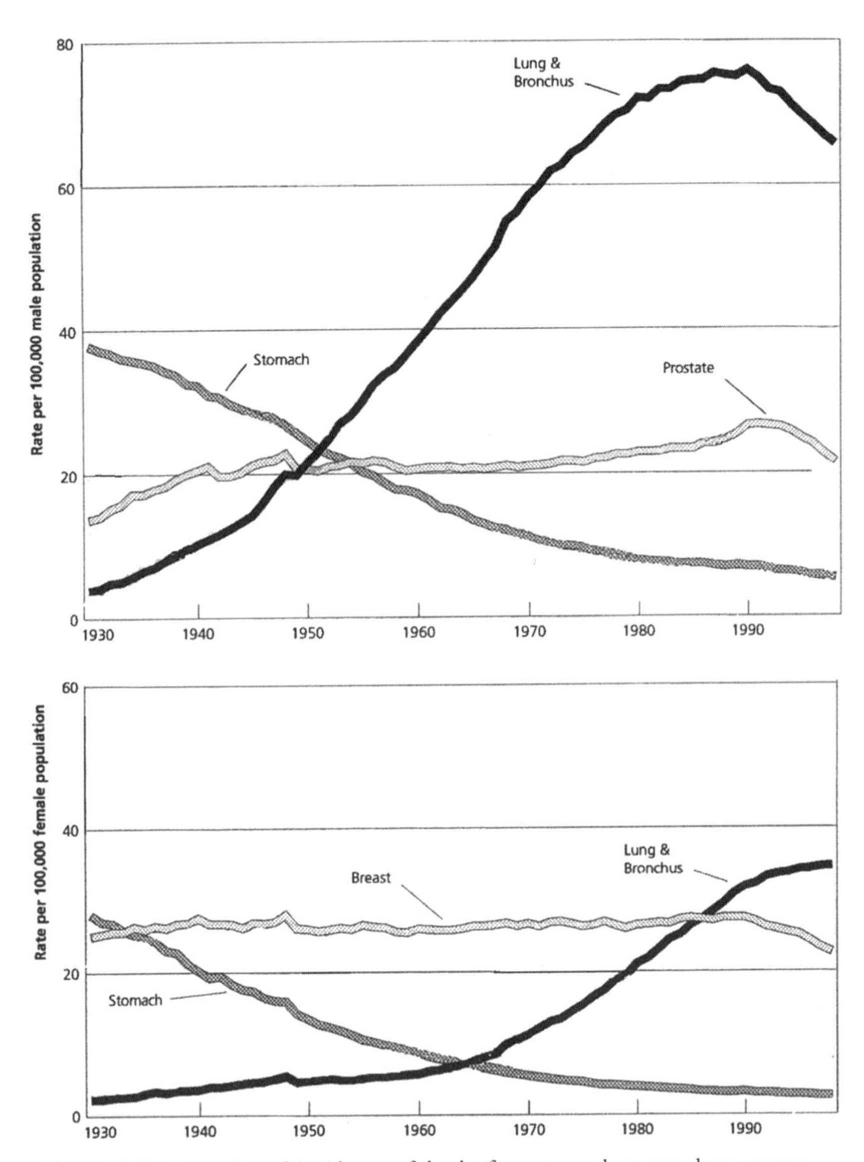

Figure 7.19. Age-adjusted incidence of deaths from stomach cancer, lung cancer, and typical male (top) and female (bottom) cancers in the United States from data published by the American Cancer Society in 2003. These should be taken as indicative of trends, but care must be taken in drawing conclusions. The death rates are not the same as the rates of incidence, and reported death rates will be affected by changes in methods of diagnosis, treatment, and other factors. Reprinted by permission of the American Cancer Society, Inc.

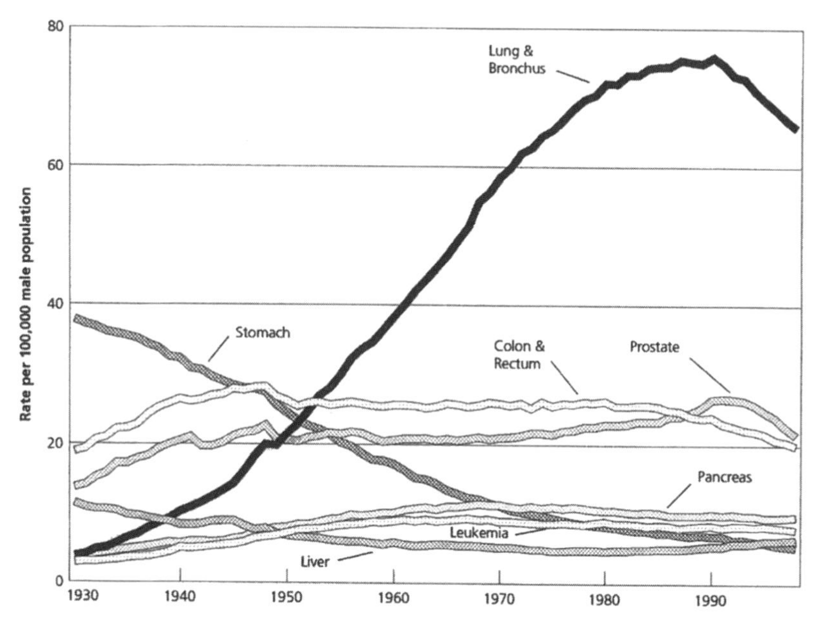

Figure 7.20. Age-adjusted male deaths from various forms of cancer in the United States from data published by the American Cancer Society, 2003. Reprinted by permission of the American Cancer Society, Inc.

finding. A reasonable interpretation might be that nitrate actually confers protection against stomach cancer, and there is now some reason to believe that a mechanism of such protection will soon be discovered.

Two further considerations put this supposed cancer risk into an even more questionable background. Nitrate is found in vegetables, even those that are grown "organically," which, as quoted above, may receive as much as 40% of their nitrogen content from anthropogenic aerial sources (that is, ammonia and oxides of nitrogen in the air), initially produced by human activity. Some vegetables contain more than a gram of nitrate in a kilogram of matter. Beetroot, spinach, lettuce, and carrots are particularly rich in nitrate, though beans, peas, and potatoes contain much less. An adult in a Western industrial society has an average daily intake of nitrate from vegetables of about 70 mg, and vegetarians considerably more.[23] The consequence is that people who eat vegetables will generally ingest much more nitrate from their diet than from their drinking water.

Someone in a Western developed society drinking water containing a high level of nitrate (say 50 mg/L) will, on average, take in another 70 mg of nitrate per day. The amount of endogenous nitrate (nitrate produced naturally within the body) generated daily by a healthy adult is 30–60 mg, of the same order as the intake from drinking water! Water is not a major source of nitrate for most humans. Statistical analysis seems to show that the cancer risk arising from coffee is greater than that from most synthetic insecticides, though less than that from the average consumption of beer.[24] It would appear that though nitrate and/or nitrite may induce some cases of stomach cancer, the principal causes are to be found elsewhere.

Other Suggested Health Implications

There are several other health effects that have been linked to nitrate.[25] Some of these would certainly be harmful and require action were they to be confirmed. For example, it has been suggested that the methaemoglobin content of a human mother's blood increases during pregnancy and that this carries a risk for the foetus. Attempts to confirm this have proved negative. There are some suggestions that nitrate may affect the DNA of rats and children, but again confirmation is lacking. Nitrate has been linked to congenital malformations, but this was not confirmed by later studies. Nitrate has been associated with increased size of the thyroid gland, with onset of hypertension, and with childhood diabetes. The problem with these rather infrequent reports is often that nitrate is selected as a critical factor in a situation where there are many variables, some or all of which could be the determining factor. It requires a lot of evidence, so far lacking, to exclude them. As with the two major dangers ascribed to nitrate, methaemoglobinaemia and stomach cancer, study of the available evidence appears to discount it as an important causative factor.[17]

What is more surprising than these doubts about the supposed dangers of nitrate is the fact that there are now claims that high levels of nitrate are actually beneficial. In any case, we cannot escape the influence of nitrate altogether because it is actually endogenous in humans.

The basis of the claims for the beneficial effects of nitrate arises from the same reaction that has been identified as a danger—the conversion of nitrate into nitrite—but accompanied by its further conversion to nitrogen(II) oxide, NO.[17] It is well-established that nitrite under acid conditions is an effective microbiocide and antifungal agent. The same applies to nitrogen(II) oxide. Whereas the precise details of action are not known, nitrite has been shown to destroy the bacteria *Salmonella typhurium*, *Yersinia enterocolitica*, and *E. coli* and the fungus *Candida albicans*, which can cause thrush. The antibacterial benefits of nitrate/nitrite have been discussed in relation to the skin, the respiratory tract, and the urinary tract. In addition, there is evidence, some from animal studies, that nitrate and nitrite can reduce blood pressure and reduce thickening of the arteries often observed in elderly people. Thus, there may be benefits in respect of heart disease. The mechanism, if such effects exist, may also involve nitrogen(II) oxide (nitric oxide), which plays a vital part in signalling within the mammalian body and is heavily involved with muscle control. Ironically, if vegetarians are proven to be less prone to gastrointestinal and heart problems, this may be due to their usually high intake of nitrate rather than to the absence of meat from their diet, but this remains to be established definitively.

Conclusions: The Role of Nitrogen in the Modern World

What arises from all these discussions is that the environment and biological systems are much more complex than single-issue campaigns might imply. In every case, the concepts of risk assessment need to be applied and judged against the expected benefit. This has not often happened. Of course, where no clear conclusion can be drawn, then the injunction "if in doubt, don't do it" should be seriously considered. However, there is little value in wasting resources in trying to reach a position of absolute safety, which is probably unattainable, when treating another problem might offer much larger rewards for the same expenditure. Nitrate seems to have been quite unfairly stigmatised. Pesticides might be another area worth a sceptical reconsideration. We need to know as facts, not simply as guesses, what would be the best productivity attainable for agricultural systems that are not to be supported by artificial fertilisers. We must determine what extra resources of manpower and land would need to be exploited in order to feed the world's population. We need to be sure that we have methods that can produce enough food without the use of pesticides. We also need to evaluate objectively every new technology, including genetic manipulation (GM) to establish the conditions under which it can be safely used, if at all. This may entail the imposition of political and economic controls of various kinds. The major political question is: Who is going to evaluate and impose these controls?

All this is not really optional in the opinion of many. The development of synthetic and biological nitrogen fixation has given us the opportunity to feed the world population, but this needs to be worked out very carefully. From the point of view of the over-supplied Western world, this challenge may not seem to be a problem. From anywhere outside this fortunate area, it is urgent. Otherwise, we are no better than was Crookes when musing about the future of "the great Caucasian race" threatened by races "to whom wheaten bread is not the staff of life." The difference is that he was a man of his time, and an imperialist, and the basis of his ideas was characteristic of that time. We cannot claim any such excuses.

Thus, nitrogen gives us the opportunity to benefit the whole of humanity. It is accompanied by real risks of eutrophication of groundwater, surface waters, and seawaters, but these should be manageable. The supposed risks of methaemoglobinaemia and stomach cancer are far from proven and may, in fact, be non-existent. Unfortunately, a population assaulted by changes it does not truly understand and cannot evaluate remains very suspicious, even though suspicion may indeed be unjustified. The motives of politicians (generally not scientists) and commercial companies (whose principal motive is profit) are not to be trusted. There is a need for people to be much more educated scientifically, so they can make their own informed judgements. The suspicion of science and scientists is also justified in many cases since, like the alchemists, so many are employed by the modern equivalents of rich princes hoping to find the philosopher's stone. The more that ordinary people know about science and subjects such as risk assessment, the better. There can be no absolutely risk-free course of action.

In modern industrialised societies, we cannot expect people to raise their own food. Most food needs to be produced far from where it is consumed. Transport and processing of food seems to be necessary. However, it is not obvious that the current organisation of the food supply, with intensive chemical agriculture, extensive food processing, and large chains of supermarkets, is the most desirable, even if it is regarded as efficient by established criteria. The questions of who controls the food we eat and who effectively decides what we eat need to be considered.

Even if we accept that some kind of intensive agriculture must be employed, along with plentiful nitrogenous fertiliser, natural or artificial, we need then to evaluate the risks of the technology we employ and decide whether the benefits discussed above merit the drawbacks involved. This is not trivial, but it seems clear that the consequences of high levels of nitrate in waters in temperate regions, for example, with respect to eutrophication, have been much over-emphasised. Possibly the situation in the tropics is different. Certainly, nitrate is not a poison in the way that cyanide is a poison.[17] Oral doses of the order of 7 grammes do not appear to do any harm, though they may cause vomiting. Ingestion of gramme quantities of nitrate over considerable periods does no short-term harm. However, reports such as these need confirmation. Nitrite appears to cause rather more adverse reactions than nitrate, but nitrite can be administered as an antidote in cases of cyanide poisoning.

Even more significant, nitrate (and nitrite) is a part of normal human metabolism, and concentrations of the order of milligrammes per litre are found in body fluids such as blood, sweat, saliva, and mother's milk. The levels are higher in urine, and oral doses of nitrate seem to be excreted rapidly, generally within 24 hours. The levels of nitrate in drinking water recommended by the European Union and the World Health Organisation are those below which no harm has been reported, and they provide a considerable margin of safety. They do not imply that concentrations of nitrate above these levels are dangerous. Large amounts of money have been invested to achieve these levels of nitrate in drinking water. Again this is a question of assessment of risk, perceived and actual.

The same applies to the risk of cancer. The rates of stomach cancer are dropping in Europe and in the United States. In any case, the most common cancers in Western societies are not stomach cancers. More generally, in some countries there has been shown to be a positive correlation between nitrate in water or body fluids and stomach cancer, but other studies show a negative correlation, and most show no correlation at all. The actual position is therefore unclear. Is this a case where the precautionary principle should be applied, and, if so, how? The answer is not obvious.

In the United States, the highest rates of death from cancer involve cancers of the lungs and bronchi, followed by breast and prostate cancers, colon and rectum cancers, and then stomach cancers (figure 7.20). The causative agents are probably all different, and some cancers may be genetic in origin. The death rates from lung cancers in the United States exceed each of the others by a factor of two or more. Tobacco is a far greater threat to human health than nitrate, and, unlike nitrate, it does not help feed anyone except, indirectly, those involved in growing tobacco and producing and selling cigarettes.

As discussed above, there is even evidence, so far not generally known or accepted or definitively proven, that suggests that nitrate may indeed be beneficial to humans. Undoubtedly, some of the products of nitrate metabolism, nitrite and oxides of nitrogen, are involved. These would be expected to have antifungal and antibacterial properties, and these may be useful in some areas of the body but unhelpful in others. The overall balance of the effects of nitrogen and its compounds on humankind is not yet absolutely clear, but currently it seems overwhelmingly positive.

Conclusions: The Role of Nitrogen in the Human Story

A review of all the history presented in this book shows how nitrogen and nitrate, directly and indirectly, have determined in part the way civilisation has developed. It reveals how we have been able gradually to understand how dinitrogen is converted from a seemingly inert gas to a vital nutrient. In addition, it carries a warning. Whenever we interfere with the environment, and this we do simply by existing, then we change the course of nature. The question to consider is whether or not the changes we induce are tolerable for the whole human race and also for the organisms with which we share the planet.

Specifically, as far as nitrogen is concerned, industrial fixation allows us to feed as many people as the human race is likely to produce, at least as long as the climate pattern does not change appreciably. There may be immediate drawbacks in the form of pollution problems in groundwaters and in consequences for human health. As discussed, these are not necessarily significant, and much more objective work is required to establish what the situation really is as far as such dangers are concerned.

There are also other implications that are outside the direct concerns of this book. Is it really desirable in the long term that the world population should stabilise at about 10,000 million? Is such a large number of people desirable? If not, how should the population be controlled? Who will make the decisions in this area, and on what basis? These questions are social, political, and economic and will require a world consensus if they are to be answered satisfactorily.

Notwithstanding, the history of nitrogen is an epic tale of human endeavour. It shows how humans have adapted their models of the world in order to confront the agricultural problems that faced them. Sometimes it seems that our ambitions have overstepped the boundaries of what was achievable. Nevertheless, the problem of feeding the human race is really solved. This has required us to develop an apparently more realistic appreciation of how the universe functions and of the nature of the chemical elements. We have had to understand much about the reactivity and nature of the elements, and from that understanding has developed the modern chemical industry of which nitrogen fixation has been such an important part. We have started to ferret out some of the deepest secrets of biology, though this quest is far from complete.

This entire endeavour is a monument to the individuals and the generations, largely unrecognised, whose hard work and intelligence ensured that much of humanity was fed. It is also a memorial to the more recent science and the scientists since the seventeenth century whose inquisitiveness, imagination, and application have placed the ultimate solution to the age-old problem of feeding the population of the world in our hands. Whether the human race will be able to act suitably in politics, economics, and sociology to use that solution is quite another matter.

NOTES

Chapter 1

1. Roger Bacon (1214–1294) was an alchemist and mathematician who worked for much of his life in Oxford and is credited both with discovering gunpowder independently of the Chinese and describing flying machines. He encouraged princes to enlist the help of science so that they could see and destroy their enemies at a distance. Similar arguments for financial support of research are still heard in the twenty-first century.
2. An amusing and learned account of Rudolf and of many other characters involved in the early development of chemistry, though with an unusual Scottish emphasis, can be found in J. Read, *Humour and Humanism in Chemistry*, G. Bell & Sons, London, 1947. A further excellent account of early chemistry is B. Jaffe, *Crucibles*, Jarrolds, London, 1934.
3. Newton's genius was widely recognised throughout Western Europe at the time, and his interest in alchemy was not then considered eccentric. See *Humour and Humanism in Chemistry*, cited in note 2.
4. The story is told in *Humour and Humanism in Chemistry*, cited in note 2.
5. The history of European porcelain is recounted in Janet Gleeson, *The Arcanum: The Extraordinary True Story of European Porcelain*, Bantam Books, London, 1999.
6. Basic chemistry textbooks contain the material presented here and also much more. Interested readers are encouraged to read more widely, but this is not necessary for the purposes of this book.
7. Presidential Address of Sir William Crookes to the British Association for the Advancement of Science, 1898. This oft-quoted speech is not easily obtainable in full. I thank the British Association for the supply of a copy of the original printed version. A further source of information is Sir William Crookes, *The Wheat Problem*, 2nd edition, The Chemical News Office, London, 1905.

8. E. E. Fournier D'Albe, *The Life of Sir Willliam Crookes*, T. Fisher Unwin, London, 1923.
9. Presidential Address of John Tyndall to the British Association for the Advancement of Science, 1874.
10. Interested readers will find plenty of material related to this matter on the World Wide Web.
11. R. R. Eady in *The Prokaryotes*, 2nd edition, chapter 22, A. Balows, H. G. Trüper, M. Dworkin, W. Harder, and K.-H. Schleifer, Springer-Verlag, Berlin, 1991.

Chapter 2

1. There are many accounts of how agriculture began to develop. A good example is Bruce D. Smith, *The Emergence of Agriculture*, Scientific American Library, New York, 1995.
2. The impact of the Mongols upon the Chinese empire is described in Yong Yap Cottrell and A. Cottrell, *The Early Civilisation of China*, Book Club Associates, London, 1975. The Mongols under Ghengis Khan also proved rather formidable foes in Western Europe. However, in Central Asia, Ghengis Khan is still regarded as a hero rather than as the rabid despoiler remembered by the Europeans.
3. E. F. Castetter and W. H. Bell, *Yuman Indian Agriculture: Primitive Subsistence on the Lower Colorado and Gila Rivers*, University of New Mexico Press, Albuquerque, N. Mex., 1951.
4. See, for example, A. Matos Moctezuma, *The Great Temple of the Aztecs*, Thames & Hudson, London, 1988.
5. J. S. Henderson, *The World of the Ancient Maya*, John Murray, London, 1998; R. J. Sharer, *The Ancient Maya*, 5th edition, Stanford University Press, Stanford, Calif., 1994.
6. J. L. Stephens, *Incidents of Travel in Central America, Chiapas and Yucatan*, Harper and Brothers, New York, 1841, reprinted by Dover, New York, 1963; J. L. Stephens, *Incidents of Travel in Yucatan*, Harper and Brothers, New York, 1843, reprinted by Dover, New York, 1969. These books contain a series of beautiful engravings by F. Catherwood, who carried a camera obscura throughout Central America. Some of the buildings and artifacts he portrays have since disappeared.
7. M. D. Coe, *Breaking the Maya Code*, Thames and Hudson, London, 1992.
8. Diego de Landa, *Yucatan before and after the Conquest* (originally *Relacion de las cosas de Yucatan* and published in 1566), translated by William Gates, republication by Dover, New York, 1978.
9. I am indebted to Professor T. P. Culbert, University of Arizona, and to Dr. L. Grazioso S., National Autonomous University of Mexico, for assistance with this discussion. See T. P. Culbert, "The Collapse of a Civilization," in *Die Welt der Maya*, E. Eggebrecht, A. Eggebrecht, and N. Grube, editors, Verlag Philipp von Zabern, Mainz, 1992, and also B. L. Turner, *Prehistoric Intensive Agriculture of the Mayan Lowlands*, *Science*, 185: 118–124 (1974).
10. F. H. King, *Farmers of Forty Centuries or Permanent Agriculture in China, Korea and Japan*, Mrs. F. H. King, Madison, Wis., 1911.
11. The reader is referred to note 2 for a clear and direct account of Chinese history. For detailed descriptions of science and technology, including agriculture, the classical source is Joseph Needham, editor, *Science and Civilisation in China.*

For this book, the particular area of interest is covered by F. Bray, "Agriculture," in *Science and Civilisation in China*, Joseph Needham, editor, volume 6, part 2, Cambridge University Press, London, 1984.
12. Theophrastus, *Enquiry into Plants*, translated by A. Holt, William Heinemann, London, and Harvard University Press, Cambridge, Mass., 1961.
13. Marcus Porcius Cato, *On Agriculture*, and Marcus Terentius Varro, *On Agriculture*, translated by W. D. Hooper, revised by H. B. Ash, William Heinemann, London, and Harvard University Press, Cambridge, Mass., 1960.
14. M. T. Varro, *On Farming*, translated by L. Storr-Best, G. Bell & Son, London, 1912.
15. Pliny, *Natural History*, volume 5 (books 17–19), translated by H. Rackam, William Heinemann, London, and Harvard University Press, Cambridge, Mass., 1961; Selections from *The Natural History of the World, commonly called The Natural History of C. Plinius Secundus*, translated by Philemon Holland, Centaur Press, London, 1962.
16. Lucius Junius Moderatus Columella, *On Agriculture*, volume 1, translated by H. B. Ash, William Heinemann, London, and Harvard University Press, Cambridge, Mass., 1960.
17. Kevin Greene, *The Archaeology of the Roman Economy*, Batsford, London, 1979.

Chapter 3

1. K. D. White, *Roman Farming*, Thames & Hudson, London, 1970.
2. R. Prothero (otherwise known as Lord Ernle), *English Farming, Past and Present*, 6th edition, Heinemann Educational Books and Frank Cass & Co., London, 1961; J. Orr, *A Short History of British Agriculture*, Oxford University Press, London, 1922. A broader, more general account can be found in G. E. Fussell, *Farming Technique from Prehistoric to Modern Times*, Pergamon Press, Oxford, 1965. The early development of fertilisers is described by G. E. Fussell, *Crop Nutrition, Science and Practice before Liebig*, Coronado Press, Lawrence, Kan., 1971.
3. There are many available versions of William Langland's poem, *William's Vision of Piers the Plowman*. One example is *Piers the Ploughman, William Langland*, translated into modern English by J. F. Goodridge, Penguin Books, Harmondsworth, Middlesex, 1968.
4. There are several treatments of this document available. However, the original, being essentially a census, is rather boring, and, in any case, was written in Latin. One of the best sources is a series that treats the English counties separately, with parallel English and Latin texts. It is J. Morris, editor, *The Domesday Book*, Phillimore, London and Chichester, 1976.
5. At Laxton in Nottinghamshire, England, the old mediaeval field system is still worked by local farmers. It is a historic remnant that typifies the working methods of perhaps eight hundred years ago. The areas worked are now protected, and they are likely to be maintained indefinitely. The whole area was mapped in about 1635, and a representation of part of this map is shown in figure 3.4.
6. There is a multitude of books covering this period of English history. A lot of information can also be found on the World Wide Web. Anyone who looks for information will be amply rewarded but should treat what is found with some circumspection. Historians no less than others have their views coloured by prejudices and misconceptions.
7. W. Morris, *A Dream of John Ball*, Longmans, Green and Co., London, 1918.

8. E. Lamond (translator), *Walter of Henley's Husbandry*, Longmans, Green and Co., London, 1890.
9. Thomas Tusser, *Five Hundred Points of Good Husbandry*, Oxford University Press, Oxford, 1984; also Thomas Tusser, *Five Hundred Pointes of Good Husbandrie*, English Dialect Society and Trübner & Co., Ludgate Hill, London, 1878.
10. Master Fitzherbert, *The Book of Husbandry*, edited by W. W. Skeat from the original edition of 1534, English Dialect Society and Trübner & Co., Ludgate Hill, London, 1882.
11. This is quoted by G. Milton, *Big Chief Elizabeth*, Sceptre Paperbacks, Hodder and Stoughton, London, 2000. The book is an account of the first English settlements in North America.
12. Sir Richard Weston, *A Discourse on Husbandrie used in Brabant and Flanders; shewing the wonderfull improvement of Land there,* William du Gard, London, 1605 [?] (but probably 1650 because the book was reprinted in 1652 and 1654).
13. S. Hartlib, *Samuel Hartlib his Legacie: or Enlargement of the Discourse of Husbandry used in Brabant and Flanders*, etc., Richard Wodenothe at the Star under St. Peter's Church, Cornhill, London, 1651.
14. Bulstrode Whitelock, *Memorials of the English Affairs from the Beginning of the Reign of Charles the First to the Happy Restoration of King Charles the Second,* published first in 1682, reprinted by Oxford University Press, Oxford, 1853.
15. An account of the tree experiment is given by J. Read, *Humour and Humanism in Chemistry*, G. Bell and Sons, London, 1947.
16. John Houghton, *A Collection for Improvement of Husbandry and Trade*, reprinted by Gregg International Publishers, Farnborough, Hants, 1969.
17. *The Library of Agricultural and Horticultural Knowledge*, 3rd edition, J. Baxter, Lewes, Sussex, 1834.
18. There are many books discussing Malthus. One that presents both the original 1798 essay and the ultimate 7th edition of 1872 is T. R. Malthus, *On Population*, edited by G. Himmelfarb, Random House, New York, 1960.
19. For a sample of Franklin's views on population, see L. W. Labaree, W. J. Bell, H. C. Boatfield, and H. H. Fineman, editors, *The Papers of Benjamin Franklin*, volume 4, Yale University Press, New Haven, Conn., 1961.
20. D. H. Meadows, D. L. Meadows, J. Randers, and W. H. Behrens, *The Limits to Growth*, Earth Island, London, 1972; another edition: Pan Books, London, 1974.
21. Garcilaso de la Vega, El Inca, *Royal Commentaries of the Incas*, translated by H. J. Livermore, University of Texas Press, Austin, Tex., and London, 1966; another less accessible source of the lifestyle of the Incas is available on the World Wide Web via the Royal Museum, Copenhagen, Denmark, and was written by a native of what is now Peru, Felipe Guaman Poma de Ayala, *El Primer Nueva Corónica y Buen Gobierno*, published originally in 1613. See also Christopher W. Dilke, *Letter to a King: A Picture-History of the Inca Civilisation*, Allen and Unwin, London, 1978.
22. The story of the exploitation of gauno and nitrate has been described in various places, and it has been interpreted differently by the different participants. A selection of sources, by no means complete, includes E. Galeano, *Open Veins of Latin America*, Monthly Review Press, New York, 1997; B. Loveman, *Chile, the Legacy of Spanish Capitalism*, Oxford University Press, New York, 1988; Sir R. Marett, *Peru*, Ernest Benn, London, 1969; S. Collier and W. F. Sater, *A History*

of Chile, 1808–1994, Cambridge University Press, Cambridge, 1996. Copies of the various treaties may be obtainable from the official government records of the various countries concerned, possibly through their embassies in foreign countries.

Chapter 4

1. A. G. Debus, *The Chemical Philosophy, Paracelsian Science and Medicine in the Sixteenth and Seventeenth Centuries*, Science History Publications, New York, 1977. This book is a valuable survey of the chemical ideas of the sixteenth and seventeenth centuries. An article dealing specifically with "aerial nitre" is A. G. Debus, "The Paracelsian Aerial Nitre," *Isis*, 55, 43 (1964). The reader is also referred to relevant parts of J. R. Partington, *A History of Chemistry*, volume 2 (1961) and volume 3 (1962), Macmillan & Co., London. Another informative text is M. P. Crosland, *Historical Studies in the Language of Chemistry*, Heinemann, London, 1962.
2. A complete set of English (British) patents is available for inspection at the British Library in London. Other national libraries probably hold similar national sets, and some may even have copies of patents from other countries.
3. The full multivolume version of the *Oxford English Dictionary* contains the etymology and an account of the uses of all the listed words, including, in this context, both the words natron and nitre. A CD-ROM version of the dictionary is also available.
4. M. P. E. Berthelot, *Les Origines de L'Alchimie*, reprint of the original 1885 edition, Otto Zeller, Osnabrück, Germany, 1966.
5. B. Jaffe, *Crucibles*, Jarrolds, London, 1934.
6. Sendivogius is also discussed more humorously by J. Read, *Humour and Humanism in Chemistry*, G. Bell & Sons, London, 1947.
7. See W. H. Huffman, *Robert Fludd and the End of the Renaissance*, Routledge, London, 1988.
8. John Mayow, *Medico-Physical Works*, E. & S. Livingstone for the Alembic Club, Edinburgh and London, 1957.
9. We have not discussed Chinese ideas because they impinged upon European scientific minds only much later. Any interested reader is referred to the monumental multivolume compendium *Science and Civilisation in China*, edited by Joseph Needham and mentioned in chapter 2.
10. A. E. Clark-Kennedy, *Stephen Hales, D.D., F.R.S.*, Cambridge University Press, Cambridge, 1929.
11. A useful and comprehensive source describing the development of this chemistry is J. W. Mellor, *A Comprehensive Treatise on Inorganic and Theoretical Chemistry*, Longmans, London, 1922–1937. Of the sixteen volumes, volume VIII is particularly relevant to this chapter. A more general account of scientific developments can be found in A. R. Hall, *The Scientific Revolution 1500–1800*, Longmans Green & Co., London, 1954.
12. R. Boyle, *The Sceptical Chemist*, Everyman Library, J. M. Dent & Sons, London, 1941.
13. For information on Newton's ideas concerning "aerial nitre," see A. R. Hall, "Isaac Newton and Aerial Nitre," *Notes Rec. R. Soc. London*, 52: 51 (1998).
14. Guyton de Morveau, *J. Phys.*, 19: 312 (1782) and *Ann. Chim. Phys.*, 1: 24 (1798).

15. Much of this information is contained in J. W. Mellor, cited in note 11 above.
16. Herodotus, *The History*, volume 2, book IV, paragraphs 181–183, translated by A. D. Godley, The Loeb Classical Library, Heinemann, London, 1921, reprinted 1957.
17. Pliny, *Natural History*, book 31, paragraph 39, translated by W. H. S. Jones, The Loeb Classical Library, Heinemann, London, 1963.
18. S. Hartlib, *Samuel Hartlib his Legacie: or Enlargement of the Discourse of Husbandry used in Brabant and Flanders*, etc., Richard Wodenothe at the Star under St. Peter's Church, Cornhill, London, 1651.
19. Daniel Coxe, *Philos. Trans. R. Soc. London* 9: 150 (1674); 9: 4 (1674); 9: 169 (1674).
20. T. Thomson, *A System of Chemistry*, 6th edition, Baldwin, Craddock and Joy, London, 1820. This book went through many editions over twenty years and was sometimes revised, so comparison of different editions throws an interesting light on the development of chemical science during those years.
21. Humphrey Davy, *Elements of Agricultural Chemistry*, Longmans, London, 1813.
22. J. Marcet, *Conversations in Chemistry*, 13th edition, Longmans, London, 1837. This was the edition referred to in this discussion, and earlier editions may well be slightly different. As with all multi-edition books, it is probable that the text was revised from time to time. It is said that the reading of this book persuaded the apprentice bookbinder Michael Faraday to turn his hand to scientific research.
23. The version consulted in the preparation of this book is J. von Liebig, *Chemistry in Its Application to Agriculture and Physiology*, 2nd edition, translated by L. Playfair, Taylor and Walton, London, 1842.
24. The discussion of this part of the nitrogen-fixation story relies heavily on two excellent reviews: A. Quispel, *Nitrogen Fixation: Hundred Years After*, H. Bothe, F. J. de Bruijn, and W. E. Newton, editors, Gustav Fischer Verlag, Stuttgart, 1988; and P. S. Nutman, "Centenary Lecture," *Philos. Trans. R. Soc. London B*, 317: 69 (1987).
25. A well-presented basic summary of biological nitrogen fixation can be found in J. R. Postgate, *Nitrogen Fixation*, 3rd edition, Cambridge University Press, Cambridge, 1998.
26. H. Bortels, "Molybdan als Katalysator bei der biologischen Stickstoffbindung," *Arkiv. Mikrobiol.*, 1, 333 (1930); H. Bortels, "Über die Wirkung von Molybdän- und Vanadiumdüngungen auf Leguminosen," *Arkiv. Mikrobiol.*, 7: 13 (1936).

Chapter 5

1. Sir William Crookes, Presidential Address to the British Association for the Advancement of Science, 1898; W. Crookes, *The Wheat Problem*, Chemical News Office, London, 1905.
2. J. Knox, *The Fixation of Atmospheric Nitrogen*, Gurney & Jackson, London, 1914.
3. H. J. M. Creighton, *J. Franklin Inst.*, 187: 377, 599, and 705 (1919).
4. J. W. Mellor, *A Comprehensive Treatise on Inorganic and Theoretical Chemistry*, sixteen volumes, Longmans, London, 1922–1937. Volume VIII is particularly relevant to this chapter.
5. H. Clement, *The Nitrogen Fix*, Ace Science Fiction Books, New York, 1980.
6. British patents may be consulted via the British Library in London.
7. Sir William Crookes, *Chem. News*, 65: 301 (1892).

8. A very detailed account of the genesis of the Norwegian arc process is given in *La Societé Norvégienne de l'Azote*, Imprimerie Hemmerlé, Petit & CIE, Paris, 1957. This has been used extensively in writing this section of the story.
9. A. A. Breneman, "The Fixation of Atmospheric Nitrogen," *J. Am. Chem. Soc.*, 11: 3 and 31 (1889).
10. These old accounts can be difficult to track down. There was the possibility that the cyanide was originally contained, at least in part, in the coal. The episode of the cyanide was described by T. Clarke in *Poggendorfs Ann.*, 40: 315 (1837), and discussed in *Dingl. Pol. J.*, 65: 466 (1837).
11. R. H. Bunsen and L. Playfair, "Report on the gases evolved from Iron Furnaces, with reference to the Theory of the Smelting of Iron," Report to the British Association for the Advancement of Science, 1845.
12. F. Margueritte and M. de Sourdeval, *Ber.*, 79 (1878); F. Margueritte and M. de Sourdeval, *C. R.*, 50: 1160 (1860).
13. This research extended over many years, and much of it is reported in patent literature—for example, that aimed at establishing the optimum conditions for dinitrogen uptake by heated carbides. For a selection of later publications, see A. Frank, *Z. angew. Chem.*, 16: 536 (1903); A Frank, *Z. angew. Chem.*, 18: 1734 (1905); A. Frank, *Z. angew. Chem.*, 19: 835 (1906); A. Frank, *J. Soc. Chem. Ind.*, 27: 1093 (1908); N. Caro, *Z. angew. Chem.*, 22: 1178 (1909).

    The best way to obtain more detailed information is to consult reviews of the period. A selection of relevant publications concerned with the development of the chemical industry includes B. Waeser, *The Atmospheric Nitrogen Industry*, volume II, J. & A. Churchill, London, 1926; J. Knox, *The Fixation of Atmospheric Nitrogen*, Gurney & Jackson, London, 1914; H. A. Curtis, editor, *Fixed Nitrogen*, The Chemical Catalog Company, New York, 1932; A. J. Ihde, *The Development of Modern Chemistry*, Harper and Row, New York, 1964; L. F. Haber, *The Chemical Industry 1800–1930*, Clarendon Press, Oxford, 1971; and D. W. F. Hardie and J. D. Pratt, *A History of the Modern Chemical Industry*, Pergamon Press, Oxford, 1966. As their titles imply, these cover the whole of the fixation industry and not just the cyanamide process.
14. See note 4, specifically p. 148 ff., for a summary of this kind of work.
15. F. Haber, 1918 Nobel Laureate for chemistry, in *Nobel Lectures 1901–1921, Chemistry*, Elsevier, Amsterdam, 1966.
16. There are several accounts of Haber's rather sad life. One of the most accessible is contained in V. Smil, *Enriching the Earth*, MIT Press, Cambridge, Mass., 2001; see also a large range of publications, including A. Cottey, *Chem. Indus.*, March 16, 1992, p. 219; R. Hofmann and P. Laszlo, *Angew. Chem. Int. Ed.*, 40: 4599 (2001); and F. Haber, *Fünf Vorträge*, Verlag von Julius Springer, Berlin, 1924.
17. A photograph of this occasion can be found at http://www.chemheritage.org. An account of the occasion can be found in S. Garfield, *Mauve: How One Man Invented a Color that Changed the World*, W. W. Norton, New York, 2001.
18. F. Haber and R. Le Rossignol, "Über die technische Darstellung von Ammoniak aus Elementen", *Z. Elektrochem.*, 19: 53 (1913). This was actually published in January of that year. A detailed summary in English was also published in 1913 in *J. Soc. of Chem. Indus.*, 32: 134 (1913).
19. The reader is referred to the publications cited in note 13 for more information on the development of the cyanamide process.

20. Carl Bosch, 1931 joint Nobel Laureate for chemistry, in *Nobel Lectures 1922–1941, Chemistry*, Elsevier, Amsterdam, 1966.
21. Some of the reviews cited above also cover the genesis of the Haber–Bosch process. However, the Haber–Bosch process, unlike the Norwegian arc and cyanamide processes, is still in operation today. Consequently, there is a much larger literature relating to it. Sources of information include A. Travis, *Chem. Indus.*: 581 (1993); M. Appl, "The Haber–Bosch Heritage: The Ammonia Production Technology", Report to the 50th Anniversary Meeting of the IFA Technical Conference, Sevilla, Spain, 1997; B. Timm, *Chem. Ing. Tech.*, 35: 817 (1963); V. Smil, *Enriching the Earth*, MIT Press, Cambridge, Mass., 2001; B. Timm, M. Appl, W. Kost, and K.-H. Tillmann, *Carl Bosch und das Ammoniak*, BASF information booklet 8/74, BASF, Ludwigshafen, 1974; T. I. Williams, editor, *A History of Technology*, volume IV, part 1, Clarendon Press, Oxford, 1978.
22. Bosch's own account given in note 20 is clear and comprehensive.
23. L. F. Haber, *The Chemical Industry 1800–1930*, Clarendon Press, Oxford, 1971.
24. W. J. Reader, *Imperial Chemical Industries: A History*, two volumes, Oxford University Press, London, 1970.
25. V. E. Parke, *Billingham—The First Ten Years*, Imperial Chemical Industries, Billingham, Co. Durham, United Kingdom, 1957; H. A. Humphrey, *Chem. Trade J. Chem. Eng.*, 105: 327 (1939).
26. A. Cottrell, *The Manufacture of Nitric Acid and Nitrates*, Gurney & Jackson, London, 1923.
27. R. W. Treharne, D. R. Moles, M. R. Bruce, and C. K. McKibben, Proceedings of the International Solar Energy Meeting, Atlanta, Georgia, 1979.

Chapter 6

The following references constitute a selection of key publications concerning the major developments presented in this chapter. They are neither complete nor necessarily the latest information on the subjects discussed. Nevertheless, they give a balanced account of the research as it appeared midway through 2003. For more detailed information, the reader should undertake specific literature searches.

1. F. Haber, 1918 Nobel Laureate for Chemistry, in: *Nobel Lectures, 1901–1921, Chemistry*, Elsevier, Amsterdam, 1966.
2. D. A. Meadows, D. L. Meadows, and J. Randers, *The Limits to Growth*, Earth Island, London, 1972. This has since been updated: D. A. Meadows, D. L. Meadows, and J. Randers, *Beyond the Limits*, Earthscan Publications, London, 1992.
3. J. E. Carnahan, L. E. Mortenson, H. F. Mower, and J. E. Castle, *Biochim. Biophys. Acta*, 44: 520 (1960) (the first report of cell-free extracts of a nitrogenase, from *Clostridium pasteurianum*).
4. H. Bortels, "Molybdan als Katalysator bei der biologischen Stickstoffbindung", *Arkiv. Mikrobiol.*, 1: 333 (1930); H. Bortels, "Über die Wirkung von Molybdän- und Vanadiumdüngungen auf Leguminosen", *Arkiv. Mikrobiol.*, 7: 13 (1936) (the demonstration on whole plants and organisms of the requirement for molybdenum, or in its absence, vanadium, to support nitrogen fixation in vivo).
5. A detailed account of the state of research related to nitrogen fixation, both chemically and biologically, can be found in G. J. Leigh, editor, *Nitrogen Fixation at the Millennium*, Elsevier, Amsterdam, 2002.
6. D. J. Lowe and R. N. F. Thorneley, *Biochem. J.*, 224: 877 (1984); R N. F. Thorneley and D. J. Lowe, in T. G. Spiro, editor, *Molybdenum Enzymes*, John

Wiley, New York, 1985, p. 221. (This work presents an essentially mathematical model of the interaction of nitrogenase with dinitrogen that is still basically accepted. However, it is interpreted in terms of specific steps of single electron transfer, eight such transfers being required to reduce one molecule of dinitrogen.)

7. R. A. Dixon and J. R. Postgate, *Nature*, 234: 47 (1972). (The knowledge that all the information necessary to construct a biological nitrogen-fixing system was localised on a relatively short length of DNA opened the way to map and study the individual genes involved. Modern technology now makes such analysis relatively easy. Thirty years ago, it was a considerable undertaking.)
8. P. E. Bishop, D. M. L. Jarlenski, and D. R. Heatherington, *Proc. Natl. Acad. Sci. USA*, 77: 7342 (1980) and later papers. (This discovery was greeted with considerable scepticism. It was only after it had been shown that a mutant bacterium that did not have the genes to process molybdenum could still fix nitrogen that the most reluctant accepted the reality of an alternative nitrogenase.)
9. J. Kim and D. C. Rees, *Science*, 257: 1677 (1992); J. Kim and D. C. Rees, *Nature*, 360: 553 (1992) (These papers described for the first time the structure of *Azotobacter vinelandii* dinitrogenase. The resolution was relatively low, and there one or two areas of the molecule that were not properly defined. However, subsequent investigations clarified them, though there were still surprises to come. The structures of several dinitrogenases and nitrogenase reductases are now known; even the structure of the usually transient adduct of these two proteins during electron transfer has been determined.) J. T. Bolin reported a dinitrogenase structure independently, and almost at the same time as Kim and Rees.
10. O. Einsle, A. Tezcan, S. L. A. Andrade, B. Schmid, M. Yoshida, J. B. Howard, and D. C. Rees, *Science*, 297: 1696 (2002) (the latest and best-resolved dinitrogenase structure, with the mysterious light atom at its centre).
11. M. E. Volpin and V. B. Shur, *Dokl. Akad. Nauk SSSR*, 156: 1102 (1964) (the first reported reaction of dinitrogen under relatively mild conditions; the detailed reaction mechanism is still to be determined).
12. A. D. Allen and C. V. Senoff, *Chem. Commun.* 621 (1965). (This paper covers the first dinitrogen complex, of ruthenium. Like many discoveries, this was accidental. The authors were trying to make an established compound by a published route but noticed that the spectroscopic properties of their product were not as expected. They did not dismiss the discrepancy but found the explanation, the first example of a new class of complex. Apparently, they had some trouble convincing the referees of the manuscript that they were correct.)
13. Some specimen publications of the time are cited here: J. P. Collman and J. W. Kang, *J. Am. Chem. Soc.*, 88: 3459 (1966) (an iridium–dinitrogen complex); A. Sacco and M. Rossi, *Chem. Commun.* 316 (1967) (a cobalt–dinitrogen complex); A. Yamamoto, S. Kitazume, L. S. Pu, and S. Ikeda, *Chem. Commun.* 79 (1967) (another cobalt–dinitrogen complex); J. H. Enemark, B. R. Davis, J. A. McGinnety, and J. A. Ibers, *Chem. Commun.* 96 (1968) (the structure of a cobalt–dinitrogen complex).
14. D. E. Harrison, E. Weissberger, and H. Taube, *Science*, 159: 320 (1968) (a complex with dinitrogen bridging between two ruthenium atoms).
15. Sacco and Rossi, cited in note 13 above.

16. M. Hidai, K. Tominari, Y. Uchida, and A. Misono, *Chem. Commun.* 814 (1969) (the first molybdenum–dinitrogen complex, taken at the time to be a model for the binding of dinitrogen to molybdenum in dinitrogenase).
17. C. E. Laplaza and C. C. Cummins, *Science*, 268: 861(1995) (the first cleanly defined splitting of dinitrogen by a simple complex, of molybdenum, yielding a nitrido complex).
18. J. Chatt, A. J. Pearman, and R. L. Richards, *Nature*, 259: 204 (1976) (the protonation of coordinated dinitrogen to give ammonia).
19. C. J. Pickett and J. Talarmin, *Nature*, 317: 652 (1985) (an electrochemical cyclic system for reducing dinitrogen, based upon stable dinitrogen complexes).
20. A. A. Yandulov and R. R. Schrock, *Science*, 301: 76 (2003) (the best chemical model yet for the active site of a molybdenum dinitrogenase).
21. K. L. Grönberg, C. A. Gormal, M. C. Durrant, B. E. Smith and R. A. Henderson, *J. Am. Chem. Soc.*, 120: 10613 (1998) (a feasible explanation of why the unusual compound homocitrate rather than any feasible alternative is bound to molybdenum in the structure of dinitrogenase).
22. M. Ribbe, D. Gadkari, and O. Meyer, *J. Biol. Chem.*, 272: 26627 (1997) (the new and unexpected dioxygen-stable nitrogenase).

Chapter 7

1. Figure 7.1 is modified from an original appearing in S. Lamb and D. Sington, *Earth Story*, BBC Books, London, 1998, p. 148.
2. Adapted from data provided by Dr. P. D. Jones, University of East Anglia, Norwich, U.K. The data represent ten-year averages for the decades beginning in the years indicated. The diagram is based upon the dataset CRUTEM2v. For a detailed discussion, see P. D. Jones and A. Moberg, *J. Climate*, 16: 206 (2003).
3. These data relate to phanerozoic times, that is, the period since green plants appeared on Earth. Plants affected carbon dioxide concentrations through photosynthesis and respiration. Changes can be estimated by isotope measurements of plant remains, rocks, and other materials.
4. Measurements of atmospheric carbon dioxide have been taken at several places around the world and using a variety of land-based and satellite measurements. There appears to be a consensus that levels are increasing. As industrialisation increases and energy use becomes more intensive, the increase can only be expected to become more rapid.
5. All the population figures are taken from the FAO statistical data bases, FAOSTAT at http:/apps.fao.org
6. V. Smil, *Global Biochem. Cycles*, 13, 647 (1999). Professor Smil has a record of publications analysing the use and effects of nitrogen. See also V. Smil, *Enriching the Earth*, MIT Press, Cambridge, Mass., 2001, and V. Smil, *Long-Range Perspectives on Inorganic Fertilizers in Global Agriculture*, 1996 Travis P. Hignett Lecture, IFDC, Muscle Shoals, Ala. This last can be downloaded from the IFDC Web site http://www.ifdc.org.
7. These data were taken from information published on the Web by the International Fertilizer Industry Association (IFA), http://www.fertilizer.org/ifa/
8. All the data for food production quoted in this chapter are taken from the United Nations Food and Agriculture Organisation (FAO) Web site, http://apps.fao.org
9. These numbers were taken from the Wisconsin Foodshed Project at www.foodshed.wisc.edu.

10. These data are quoted by O. C. Bøckman, O. Kaarstad, O. H. Lie, and I. Richards, *Agriculture and Fertilizers*, Norsk Hydro, Oslo, 1990.
11. I thank Professor A. E. Johnston for providing me with an up-to-date figure of the data from the Broadbalk experiment, which is reproduced here as figure 7.15. See also K. W. T. Goulding, N. J. Bailey, N. J. Bradbury, P. Hargreaves, M. Howe, D. V. Murphy, P. R. Poulton, and T. W. Willison, *New Phytol.*, 139: 49 (1998).
12. Note 10, p. 110. A more popular presentation, with appropriate drawings, was reported in the London newspaper *The Independent on Sunday*, March 21, 1993, which shows a beautifully symmetrical, intensively farmed carrot receiving 30% of its nitrogen from the soil, 30% from chemical fertiliser, and 40% from airborne chemicals, compared to a scrubby organic carrot that gets 60% of its nitrogen from soil and manure and 40% from the air. This rather biased presentation is probably scientifically accurate and was based upon findings from Rothamstead Experimental Station in southern England.
13. The role of nitrogen contained in organic compounds rather than in nitrate in the nitrogen cycle may have been underestimated. There are many papers on this topic. See, for example, C. L. Goodall, J. D. Aber, and W. H. McDowell, *Ecosystems*, 3: 433 (2000); S. Cornell, A. Rendell, and T. Jickells, *Nature*, 376: 243 (1995); B. Seely, K. Lajtha, and G. D. Salvucci, *Biogeochemistry*, 42: 325 (1998); and M. R. Churdhill and C. T. Driscoll, *Global Biogeochem. Cycles*, 11: 613 (1997).
14. GESAMP 1990, IMO/FAO/UNESCO/WMO/IAEA/UN/UNEP Joint Group of Experts on the Scientific Aspects of Marine Pollution, *The State of the Marine Environment*, Blackwell Scientific Publications, Oxford, 1990. For a detailed discussion of the problems in the Gulf of Mexico, go to www.csc.noaa.gov/products/gulfmex/html/rabalais.htm.
15. A good summary of many of the topics discussed in this chapter is presented by B. Lomborg, *The Skeptical Environmentalist*, Cambridge University Press, Cambridge, 2001. However, some of his conclusions are open to dispute.
16. T. M. Addiscott, A. P. Whitmore, and D. S. Powlson, *Farming, Fertilizers and the Nitrate Problem*, CAB International, Oxford, 1991.
17. J. L'hirondel and J.-L. L'hirondel, *Nitrate and Man*, CABI Publishing, Oxford, 2002.
18. M. Lægrid, O. C. Bøckman, and E. O. Kaarstad, *Agriculture, Fertilizers and the Environment*, CABI Publishing, Oxford, 1999.
19. Note 17 contains a good discussion of the incidence and causes of methaemoglobinaemia.
20. Report on WHO meeting Health Hazards from Nitrates in Drinking Water, Copenhagen, March 5–9, 1984, also quoted in note 17.
21. National Research Council, Subcommittee on Nitrate and Nitrite in Drinking Water, National Academy Press, Washington, D.C., 1995.
22. These data, and a more general discussion of other observations, are presented in note 16.
23. This is discussed in more detail in note 18.
24. B. Lomborg, Note 15, has a good discussion of the relative cancer risks of pesticides and various foods and drinks.
25. See the discussions in note 16 and especially in notes 17 and 18.

INDEX

Note: figures and tables are indicated by *f* or *t* after a page number, respectively